BLADESMITHING

A Step-by-Step Guide to Forging Your Own Knives for Beginners

By Jake Welsh

BLADESMITHING

© **Copyright 2020 - All rights reserved.**

The content contained within this book may not be reproduced, duplicated or transmitted without direct written permission from the author or the publisher.

Under no circumstances will any blame or legal responsibility be held against the publisher, or author, for any damages, reparation, or monetary loss due to the information contained within this book. Either directly or indirectly.

Legal Notice:

This book is copyright protected. This book is only for personal use. You cannot amend, distribute, sell, use, quote or paraphrase any part, or the content within this book, without the consent of the author or publisher.

Disclaimer Notice:

Please note the information contained within this document is for educational and entertainment purposes only. All effort has been executed to present accurate, up to date, and reliable, complete information. No warranties of any kind are declared or implied. Readers acknowledge that the author is not engaging in the rendering of legal, financial, medical or professional advice. The content within this book has been derived from various sources. Please consult a licensed professional before attempting any techniques outlined in this book.

By reading this document, the reader agrees that under no circumstances is the author responsible for any losses, direct or indirect, which are incurred as a result of the use of information contained within this document, including, but not limited to, — errors, omissions, or inaccuracies.

BLADESMITHING

Table of Contents

Introduction vi

Chapter One: Getting Started 1

    Picking a Space for Your Shop 3
    Setting Up Your Shop 11
    Shop Safety Tips 17

Chapter Two: Designing Your Knife 25

    The Anatomy of the Knife 27
    Popular Knife Designs 31

Chapter Three: Tools Needed 46

    The Beginner's Gear 48
    Intermediate Gear 52
    Expert Gear: Forging 57
    Extra Gear or Alternatives 62

Chapter Four: Grinding 69

    Step One: Grinding Out the Design 71
    Step Two: Finding the Center 73
    Step Three: Grinding the Edge (Pre-Heat Treatment) 75
    Step Four: Finishing the Blade (Post-Heat Treatment) ... 79
    Types of Knife Grinds 81

Chapter Five: Heating and Hardening 89

    What Equipment You Need to Heat Treat Your Blades 90
    Step One: Preparing the Forge 93
    Step Two: Heating the Blade 95
    Step Three: Quenching the Blade 98
    Step Four: Tempering the Blade 99
    Step Five: The Finishing Touches 103
    Step Six: Adding a Handle 104

Chapter Six: Creating a Sheath 112

BLADESMITHING

Step 1: Preparing the Shape 114
Step 2: Dying the Leather.. 115
Step 3: Etch Out a Groove 116
Step 4: Shape, Cut, and Glue on the Edge's Bottom 116
Step 5: Make the Belt Loop 118
Step 6: Fold, Mark, Drill, Groove, and Stitch the Sheath
Together 119
Step 7: Cut Off Excess 120
Step 8: Dye, Burn or Decorate 121
Step 9: Form the Final Shape 122
Step 10: Finish Decorating... 124

Chapter Seven: Maintaining Your Knives 129

Keep It Dry; Only Hand Wash 130
Never Put Your Knives in the Dishwasher 131
Keep It Protected in a Sheath or Wrap 132
Clean the Blade After Cutting Anything Acidic........................ 132
Oil the Blade After Cleaning 134
Sharpen the Grind Yearly (at the least).. 134
Sharpen Your Blade as Needed 135

Final Words 139

INTRODUCTION

Those of you who have picked up this book because you are looking to get into bladesmithing should already have a good idea of what this practice entails. However, those who are reading this introduction and considering if this skill is something you would like to invest in—this first section is for you.

Bladesmithing is the practice through which we craft blades. Everyone who is new to the skill may immediately bring to mind images of blacksmiths hammering red hot iron and crafting medieval weapons like swords or axes. This absolutely falls into the category of bladesmithing, as these are, without a doubt, blades. However, this understanding of the practice is limited in scope.

The art of bladesmithing is about far more than swords and weapons; it is about knives and tools. When was the last time you held a sword? Chances are never. Most people have no reason to, after all. But when was the last time you held a knife? Dinner? Lunch? Breakfast? Cutting bread? Buttering that bread? How about in the workshop? Do you have blades for cutting wood or other materials?

Many of us don't realize that we are surrounded by blades all the time. What's more, somebody had to make those. Although many are now mass produced, there is a telling lack of quality to these blades. A mass produced blade is prone to breakage, thus they need to be replaced often. This is why professional kitchen chefs will typically invest in expensive knives. They want quality, longevity, and an impressive functionality. You may think that a blade only has one function: to cut. This is mostly true, but how well it cuts and what kind of cut it makes is entirely different from blade to blade.

There is far more to bladesmithing than simply smacking hot iron against an anvil. Therefore, it is my goal with this book to introduce to you just how important and useful this skill is. It goes without saying that it can improve the quality of blades you have in your household greatly. However, even more than that, it can be used to earn money if you wish to sell your creations. On top of that, although it is a specific skill, there is a lot of crossover between bladesmithing and other forms of metalworking or even crafting in general. I am a proponent of learning at least the basics of as many crafting skills as you can because it allows you to see how you can weave together seemingly disparate skill sets to create amazing arts and crafts.

In this book, we will begin by looking at how you would get started bladesmithing. This means we'll need to set up a space in which to work. This is a little more complicated than something like crochet or even woodworking. Crochet is a crafting skill that you can take with you anywhere, and bladesmithing is certainly not so convenient. Woodworking is a little closer to bladesmithing, in that you require a workshop; however, bladesmithing is a more dangerous activity, and the amount of heat used in the process can be incredibly uncomfortable. This chapter will look at how you plan for and set up your workshop. It will also go over the more important safety tips that you should be aware of going forward.

Safety is an incredibly important part of bladesmithing, or really any form of crafting that uses dangerous equipment. Because it is so important, we will be seeing more about safety throughout the book. These tips aren't gathered into a single chapter, but rather, they are spread out, so they can be directly noted when we are discussing the relevant topic. For example, if the book was front-loaded with all the safety advice needing consideration, then it would be much easier to forget the tips that applied to grinding when we get to the appropriate chapter. Thus, we will look at safety tips when grinding a blade in the chapter in which we discuss grinding. It is my hope that this section will help make sure these tips sink in because they are integral to enjoying this skill.

Despite setting up our workshop, we won't start using it until later in the book. Chapter two is a continuation of chapter one how it is focused primarily on the planning and designing step of the process. In chapter one, you would start to design your workshop. In chapter two, would you learn all about designing your knife. You might think that you understand blades. After all, how complicated could a piece of sharp metal be? However, if you skip out on design, then you'll understand real quick that it isn't quite so simple. Consider questions, such as how long you should make the hilt or how thick the blade should be, and you'll probably start to realize that you know less about blades than you thought. Don't worry; by the end of this book, you're going to know plenty.

Chapter three still doesn't quite take us to the hands-on segment of the book, but we're getting close. This chapter looks at all the various tools that would be filling your workshop. The cool thing about bladesmithing is that there are actually way fewer tools needed than people may think. To be clear, there are a ton of tools that *can* be used in bladesmithing, but you only *need* a handful of tools to actually do the work. You can always purchase more tools down the road to make your experience easier or more enjoyable, but I always recommend starting with the minimum, and then purchasing more over time. This reduces the amount of money it takes to start bladesmithing, and that means you can start practicing your skills and selling blades sooner rather than later.

In chapter four, we'll be getting hands on, though we won't be worrying about the fire just yet. We will be starting with grinding. If you've ever watched anything with a blacksmith in it, then you've likely seen a grinding wheel before. But do you know how to use one? Don't worry—you don't need to. That's what this book is for!

Chapter five is where we finally get into what people associate most strongly with bladesmithing: heat. This chapter will focus on the heating, shaping, and hardening of your blade. All that hard work you put into designing your blade will finally pay off in this chapter.

Now that you have a finely crafted blade, you could just stick it in a drawer if you want; but, since we've gone through all the effort of designing and making it ourselves, why not create a sheath just for your blade? In chapter six, you'll do just that.

The final chapter of this book deals with maintaining your knives. It is one thing to make a knife, but it is another to keep a knife in good condition. It isn't particularly hard. In fact, it is actually shockingly easy to maintain a blade. Despite that, so many people often don't bother to learn the necessary skills to do so. However, you are different, and after reading chapter seven, you will know how to maintain your blade, so I trust that you will.

Thus, there's the roadmap. From setting up to creation to maintenance, this book will take you through the entire process of designing and crafting your first blade. So what are you waiting for? Turn the page and get ready to sharpen up your skills!

CHAPTER ONE

GETTING STARTED

The first step toward learning bladesmithing is to acquire or build a space for it. The bladesmith's shop is a lot like a blacksmith's shop: there are many of the same tools and certainly as much heat. If you are just looking to try out bladesmithing—to see if it is something that you are interested in—then I would recommend that you use Google to find a local shop and see if you can observe them at work. Many blacksmith shops even teach beginner lessons on blacksmithing, and an afternoon course could give you hands-on experience quickly.

The reason I recommend this is because bladesmithing is expensive to start. You will need to set up a space for it, and that requires commitment because you either need to convert space you already have into a shop (like a garage), or you need to rent a space. Renting a space can prove to be difficult at times because many people don't want to allow all the necessary changes to the space needed by the equipment and practice.

This chapter will assume that you plan to convert a pre-existing space into a shop, say your garage or a room in your house. This space will be filled with tools eventually, but we won't be worrying about them just yet; we'll get to the tools in chapter three. Right now, we will be focusing on the space itself, what we need to do to ensure that it functions properly, and how we can maximize our shop's layout to increase productivity. Following this, we will have a discussion about safety and set down some simple albeit crucial safety rules.

Picking a Space for Your Shop

Picking a space for your shop is often a matter of convenience. If you already have a garage, then chances are, you will end up using your garage. If you have an extra bedroom that isn't being used, then you may decide to set your shop up there. Regardless of the rest of the information that this section holds about your shop, it is often convenience that is the first and most important factor in choosing a location for many people. This is understandably so, too, when you consider how much effort it takes to arrange for a space beyond your own property.

However, despite the convenience, there are some major factors that you need to weigh. These considerations can help you to pick the right space, but I don't want you to feel limited by them. For example, if you decide to build your shop in your garage because that is what's available, then you shouldn't panic if you realize you don't have adequate ventilation in place. While this is an important part of your shop, which we'll discuss in depth in a few moments, it isn't something that you have to look for to begin with. If you are willing to invest money into your shop, then you can add ventilation, electricity, along with whatever you need, so long as you are willing to spend the money and time to deploy it.

Yet, it must be noted that the more of these requirements you can check off before altering your space, the easier you will find the whole experience of setting up your bladesmithing shop. It will already end up taking some time and money, thanks to the tools, benches, and all the other necessary or recommended resources that you must acquire and place. I wouldn't expect to have a bladesmithing shop up and running in less than a month. Some people can prove this likely outcome wrong, especially those who are willing to spend lots of money at once for a head start. However, it is more realistic to think of setting up your shop as a long-term project. We'll talk more about setting up the shop itself in just a moment; first, let's look at the requirements you need to consider for your space before deciding on a spot.

There are three elements that we must consider and prepare before moving on: ventilation, electricity, and size. The first element is one of safety, and it could be considered to be a shop safety tip, although you must understand that its importance goes beyond a mere tip, as it is closer to a feature that your shop absolutely must have. If you cannot achieve this, then your shop will never be safe. The second is a matter of practicality, whereas the third is a matter of comfort in more ways than you may think.

Most of what you will be doing in your shop is working with metal. That is the largest component of bladesmithing after all. However, that isn't the only step. Consider creating a handle for your blade, for example. This is most often done with wood or horn, though there are certainly other materials that can be used. These can create a lot of dust and debris light enough to make the air dangerous to breathe. If you've ever worked with wood before, then you'll know how important ventilation is for removing these particles from the air and keeping you safe.

But your ventilation isn't only used in these cases. This is just an easy one for people to relate to. Another major factor in this is that you are dealing with a forge. A forge can create a lot of smoke, which is not the type of the air pollutant that you want to be breathing in. However, *yet again*, this isn't the only concern. There are also poisonous gases that you must be aware of, as carbon monoxide poisoning. Depending on the type of metal you are using, there may be other chemicals that you should be worried about. Have you ever heard of metal fume fever? This is an illness that can induce chills, nausea, chest pains, headaches, fever, dizziness, coughing, and more. In addition, that's only a minor issue—further exposure to these fumes could cause shock, bloody diarrhea, rash, and just so many more uncomfortable and deadly symptoms. Metal fume fever is caused when you breathe in the byproducts caused by heating siler, gold, platinum, or stainless steel, which can release zinc oxide, aluminium oxide, or magnesium oxide.

Ensuring that you have adequate ventilation isn't overly hard. If you can pick a space with a chimney, then you are already well on your way toward setting up ventilation. If you don't have a chimney, then what about a window? If you have access to either of these, then you should set up a ventilation system to pump out the air in the shop. The best ventilation systems will also bring in clean air from outside, thereby both pumping out many of the pollutants in the air while also reducing the ratio of clean air to polluted air, which then further reduces the risk to your health.

Ventilation is such an important step for setting up your bladesmithing shop that I would recommend giving up on your space if you can't properly ventilate it. Bladesmithing can be fun, but it isn't worth dying for.

The next thing to worry about is electricity. There are a lot of tools involved in bladesmithing that will require electricity. Depending on how large your shop is and how many people are working at any given time, this could cause a lot of strain on your electrical system. Before picking your shop's spot, go through and check to see how many electrical outlets it has. You'll want to have at least two or three outlets, though you can always purchase a power bar or the like to expand the coverage.

More important than the outlets is the total amount of power the room can use. You don't want to blow a switch while working in your shop, but you want to have your ventilation system on, whatever tools you are using, and the lights. This can eat up a lot of juice. If you are unsure of how much electricity your room has, then you should purchase a monitor and use it to check. You can also use a monitor to check how much electricity your various tools eat up. A typical monitor plugs into the outlet, and then has the tool plug into the monitor. The monitor doesn't prevent the electricity from getting to the tool; rather, it works as just another conduit to pass through. However, it is a conduit that measures the amount of electricity passing through to give you an accurate rating of how much electricity the tool uses. This is a great way to get a sense of how many tools or appliances you can run simultaneously in your shop.

If you don't have enough electricity in your space, then you may want to consider changing spaces because it will be much easier to pick one with proper electricity to begin with. However, if you are stuck with this space, then there are options. The most expensive option is to hire an electrician to rewrite your electrical systems to give your shop more watts. This requires hiring a professional and having them come and check out your place, then getting to work on it. This process can take a lot of time. Another option, which is a little less expensive, is to purchase a generator that you can use to supply extra watts. This option will work in a pinch, but it isn't nearly as effective as hiring an electrician. Your final option, which is both the cheapest and the easiest, is to use extension cords to bring in power from outside of the room. This doesn't always work, but it's cheap to test out, so it is still worth testing this option first.

Electricity is clearly important because your electrical tools won't function without it, but it isn't nearly as important as ventilation. Ventilation will keep you alive, whereas electricity will keep you working.

The final piece to consider when picking your space is the size. This one is a bit complicated because it doesn't follow any simple rules like the other two components do. For example, ventilation is necessary for your health and electricity is necessary for you to work, but is size just as necessary? You can still work in a small shop—in fact, there are many who prefer a smaller shop—but a large shop will also have its advantages.

BLADESMITHING

The size of your shop is clearly important for calculating how much space you will have available for equipment. You will want to have enough space for any of your necessary gear; however, if it is too tight of a fit, then you'll find yourself having a hard time navigating the space. A larger space will be easier to move around in, and it can fit more tools. On the other hand, as a trade-off, it will also take more time to finish projects because you will be required to move from one part of the room to another more often. This may seem a minor inconvenience—and it really is—but it is one that does add up over time. You may only spend an extra ten minutes a day moving from one part of the room to the next, but in a week, that would be more than an hour wasted. In a year, that would be about 60 hours of time wasted. It adds up.

There is another issue of size that isn't so immediately apparent when first considering the heating and cooling of the space and the problems that could arise with them. When you are working the forge, it will be hot—that is expected. However, what about when you're in there working during the summer? It can broil without the forge being in use, so imagine what it would be like to use the forge in the summer. Therefore, you definitely want to get some air conditioning to cool you down. Likewise, there are some things to consider when working in the winter. If the forge is going in the winter, then the shop should be nice and toasty; otherwise, the rest of the time, you'll find it chilly. This is exactly when a heat source would be a great fit. So, how does size fit into this?

The smaller your shop is, the easier it will be to warm it up or cool it down. The larger the shop, the harder it is in either direction. Thus, there is a balancing act between size and temperature that can be truly frustrating. If you have the option, I would recommend having a heating pump installed since they are great for both heating and cooling. If you don't have that option, then you will need to use fans and heaters, and these eat up even more of your precious electricity.

I don't recommend changing your spot simply because of temperature issues. You have options for dealing with it, though you need to be aware that dealing with it through electricity will cause more stress on the system. However, if you can select a space that isn't too small and isn't too big, then you'll have a much easier time with this.

In addition, there is one last thought before we move on to setting up the shop: you should measure the door beforehand. Although bladesmithing doesn't use materials that are all that large—such as how a woodworker does so when bringing in boards—there are still tools that are quite large, and these could prove difficult to get into the shop if your entrance way is too small. Make sure to get the measurements for your door, so you never find yourself frustratingly trying to squeeze a new piece of equipment into the space.

Setting Up Your Shop

Setting up your shop isn't that difficult to do once you've gotten ventilation, electricity, and heating out of the way. As with the previous section, there are three key pieces to setting up your shop that we'll be going over, plus an exercise to ensure that you are using your space to the best of your ability. I am a strong believer in the importance of planning; I believe that you should have a plan in place for your bladesmithing shop before you purchase even the cheapest of tools. If you have a plan, you will be a thousand times more likely to succeed.

We will first be looking at the three main components for a successful shop setup. These are the division of the tools, placement of the tools in regards to workflow, and placement of tools in regards to maximizing your available space. Once you understand these aspects, you will be able to combine them with the safety tips that follow afterwards to create a productive and safe shop environment in which your bladesmithing can thrive.

We want to keep all of our tools in the same place. This doesn't mean that you will want to keep your hammer and screwdrivers on top of your belt sander or inside your forge; rather, it is best to categorize your tools and use this for sorting them. You may put the tools you use for shaping the blade in one place, those for hammering in another, the tools for making the hilt in a third place, and those for making a sheath in yet another. The categorization of your tools is entirely up to you, though I do recommend categorizing them by use. Either way, it is important that you do so to create a sense of organization.

When you have organization in your shop, it becomes much easier to work in. When you finish using a hammer, put it back where it belongs. It might be easiest to just reach over and put it down on the nearest table, but if it doesn't go there, then you absolutely shouldn't. You may think that you have a great memory and would never forget where you left something, but why risk it? Plus, what happens if you need to ask someone else to grab you a tool? By placing your tools back where they belong—in their rightful place, according to your own categorizational rules—then you will never have to deal with going on a scavenger hunt for the tool you need, when you need it. Instead, you can go to where you already know it is and find it within seconds.

I mentioned placing your hammer down on the nearest table. This serves as a good piece to branch off into the second major point, which is the maximization of your available space or where you put your tools. There will be plenty of tools that you'll be bringing into your workshop before too long, but the first piece of equipment you should buy is a workbench or table. I strongly believe that, so long as there is enough space in your shop, you should have as many tables as you can. Of course, there is a limit because you'll run out of uses for that surface space eventually, but this won't be a problem in small to medium sized shops—only in large shops could you have too many tables.

Table space is necessary for pretty much everything when it comes to crafting. Have a table where you draft up plans, another for works-in-progress, then one or two for sheathing and making hilts. The best tables for a workshop like this are those with a bottom shelf slightly above the ground. This gives you shelf space on which you can easily store more gear or supplies. If you can't get your hands on tables with this added shelf, don't worry—you can always stick a half-sized tool cabinet underneath a table to maximize that storage space.

Another piece of useful furniture for maximizing your space is shelvess. The shelves in your workshop don't need to be the same style as the shelves in your bookcase; in fact, you can actually get away with tossing out the shelves themselves and just sticking with the hooks that they rest on. Many tools, like saws, can be hung up on hooks for easy access, and many bladesmiths work a circular loop onto the end of a blade, so they can hang them up until they are ready to work with them. This is a great way of using your wall space to the best of your ability.

The final component we'll look at before our planning exercise is travel distance. We spoke about travel distance in the last section in regards to the size of your workshop, but there is more to it than that. As you spend more time in your workshop, you'll figure out which areas you move between most often. Once you have these in mind, I recommend reorganizing your workshop once again. In fact, I believe it is worth reorganizing or at least reconfirming your appreciation of the current organizing on a yearly basis. There are some things that shouldn't be moved—such as which tools are in what drawer—though many of the elements in your shop can be moved around to maximize the space and improve your experience. Rather than sticking to the first design you make, it is best to update and upgrade your layout regularly.

Speaking of layout, let's get into that exercise. For this exercise, you will need some graph paper, a pencil, and a tape measure. You will also want to read chapter three and put together a list of the tools you plan to purchase first. Once you have these, you can start the exercise.

Begin by measuring the room to get the dimensions. Use your pencil and graph paper to graph out the room according to the dimensions you just gathered. Make sure you mark the door too by first measuring it and taking note of how wide it opens, and whether it swings inwards or outwards. It is usually a smart idea to make down the dimensions on the graph paper itself, so you have them on the same page as your plan.

If you have the list of equipment you will be purchasing, then the next step will be a little bit of research. You will want to shop around for your tools if you want to find the best price; however, when you do, make sure to take note of the dimensions of each tool. This will give you the required equipment size you will need to bring into the room. You can start to add these to your diagram to see how much space they'll take up and how much room it leaves you with. Work through all of your tools, tables, and other equipment in this manner, adding them to the graph where you want to place them.

Don't expect to move through this exercise with only a single diagram. This exercise is valuable for how it lets you get a sense of your workshop before you purchase anything. However, getting your workshop right is often a matter of trial and error. Try putting your forge over in the corner—does that work? No? Perhaps the corner is better suited for a table then? Just in these two questions alone, you've drawn out, erased, and redrawn on your diagram again. After a little bit, this will start to get a bit messy to work with. So, be prepared to sketch a few diagrams before you find the one that is right for you.

Shop Safety Tips

As mentioned above, we will be looking at safety tips throughout the book because there is a lot about bladesmithing that can be dangerous. Thus, here are five safety tips that you absolutely *must* consider and have in place before stepping foot in your workshop.

Wear Protective Gear: Protective gear is an absolute *must* in your workshop. You shouldn't allow someone to enter if they aren't at least wearing steel-toed boots. We're working with metal here, and all it takes is one sharp piece falling to the ground to slice your foot clean open.

You should also wear eye protection, since you don't want hot metal getting into your eyes. Likewise, you don't want to lose your hearing when working with your louder tools, so make sure you have some ear protection handy too.

Although the ventilation of the room will help prevent poisoning, you shouldn't trust that by itself. I say this, not because it is untrustworthy, but because it is always best to be overly cautious rather than under cautious. Wearing a mask will help reduce the number of harmful pollutants that get into your lungs, especially when paired with good ventilation.

If you're working the forge, then you should get yourself a proper blacksmith's apron. These are weighted aprons that protect your body from the heat and the metal. Speaking of which, sleeves are also an important piece of protective equipment because they protect your arms, but you must be careful not to wear baggy clothing that could catch on stuff.

Finally, you will need a pair of thick gloves. We're working with blades here, and that means the risk of getting cut is astronomical compared to other forms of crafting, like woodworking. Invest in the highest quality pair of gloves you can get your hands on.

Unplug Cords That Aren't Being Used: This tip is good both for your safety and keeping your electricity bill down. When a tool is plugged in, it eats up a small amount of electricity, even if it isn't turned on. The current is still moving into the object, and this uses a minor amount of electricity that adds up over time. It also reduces the overall electricity you have available to use in your workshop.

But, more importantly, unplugging your tools ensures that you never accidentally turn them on and injure yourself. Imagine you were moving a new table into your workshop but forgot you left the saw plugged in. There is now a chance that you will bump it, fall into it, or, in some other way, accidentally activate the saw and cut yourself. This would be a horrific event to experience, especially when it can be avoided by simply unplugging your tools when you are finished using them.

Treat Everything As Hot: Okay, maybe not everything. Anything that comes into contact with the forge should be treated as if it was burning hot and would melt the skin right off your hands. The forge might not even be going, but you should still treat it in this manner. This point is just smart to learn and get into the habit of doing, so you never end up with third degree burns thanks to a moment of forgetfulness.

Listen To Your Body: Your body tells you things all the time. When you're hungry, your stomach growls; when you're tired, you yawn; when you're angry, your cheeks flush red and your stomach twists in knots. There are all sorts of signs that your body gives you to tell you what type of shape it is in, and these are all signs you need to listen to.

Working in your shop will be physically demanding. There is the physical work itself, hammering and lifting, among other tasks. However, you need to remember that it often gets quite hot, and there is a lot of information you need to process and keep an eye on while working. The heat can cause your brain to get tired before your body, and when your brain is tired, it becomes easy to ignore the signs your body is giving. You should be taking breaks at regular intervals in order to give your mind and body a rest.

However, if you ever feel dizzy, tired, hungry, or sick, then you should put down the tools and leave the workshop. You should also never enter your workshop if you have been using drugs or alcohol because your brain will be functioning at a lower level, and it will be much easier for you to ignore the warning signs indicating that there is something going wrong.

Keep First Aid On Hand: This one should be rather straightforward. Your workshop is filled with things that can cut you. This is bladesmithing after all. However, it is also filled with things that can fall on you, burn you, or damage your hearing. The pressure in the room can get quite a bit heavy and cause pain, especially if you are working in the winter and the air temperature changes drastically. There are also poisonous fumes to be worried about, other things that can stub your toes on and worse.

Your workshop is an extremely dangerous place, so always, *always* keep a first aid kit on hand. It should be on the wall, clear as day to see and easy to reach, and you should always keep it fully stocked—that means that you must replace anything you use from it. If you don't keep a safety kit in your workshop, then you shouldn't have a workshop.

Chapter Summary

- Picking the right space for your bladesmithing shop requires more than selecting the first place you see. You need to balance the needs of the workshop against the space you have.

- If all you have is a single space, then you can certainly convert it into a workshop; however, it may take more effort than it is worth. Sometimes, it is easier to rent a new space than convert one you already have.

- There are three key elements you will want your workshop to have: 1. Size. 2. Proper ventilation. 3. Enough electricity to power your tools.

- There are quite a few pieces of gear that will take up a lot of space when it comes to bladesmithing. Picking too small of a place can make it difficult to fit everything you need into the workshop. However, picking a space that is too big will make it incredibly difficult to regulate the temperature. You will want to be able to heat it up when it is cold and cool it down when it is hot.

- Working with metal can cause deadly fumes, which would poison you if you didn't have proper ventilation. You may also find yourself cutting wood to create handles or other elements related to bladesmithing, and this can fill the air with particles that would be harmful if breathed in. Getting proper ventilation in the space will ensure that you never accidentally harm your lungs.

- The tools that make up a bladesmithing shop don't start out eating up a lot of electricity; however, as the shop expands, so too will your power needs. Get yourself an electricity monitor, as doing so will allow you to plug your tools into the monitor, and then plug the monitor into the wall socket. This will let you know how much electricity your tools are taking up, and you'll be able to balance your electricity needs better this way.

- Setting up your shop is a matter of dividing the tools in a logical manner, placing the tools in a way that follows your workflow, and arranging them in a way that maximizes your available space.

- Dividing your tools should be done based on their purpose; the tools you use for cutting should be placed together; the tools for making sheaths are to be kept together; and so on. This makes it easy to know where to find any given tool you may need. Just remember to put them back when you are done, or it will defeat the purpose.

- As you get more time working in your shop, you'll get a sense of how projects flow. Do you start with cutting handles, then move onto blades? Do you start with the blade and work your way toward a knife from there? Regardless of the workflow you use, set up your shop so projects move logically from one place to the next without having to jump all over the place. This will help reduce the travel distance in the shop.

- Maximizing the space in your workshop is simple, but it will improve your experience greatly. Use shelving units, tables with shelves underneath them, and wall-mounted hooks for hanging tools and blades. By maximizing your space, you will be able to fit more tools into a small workshop and limit the clutter.

- Wear protective gear when in the workshop. Steel-toed boots, gloves, a long-sleeved shirt, a blacksmith's apron, and a mask are all important safety gear.

- Unplug cords when you finish working with a tool, so you don't accidentally activate them.

- When working with a forge, it is best to assume that every piece of equipment is incredibly hot; that way, you will never burn yourself when reaching for a tool absentmindedly and forgetting about the temperature.

- Your body will tell you when it isn't feeling well. Dizziness, thirstiness, hunger, pain, and fatigue are all ways that your body will inform you that it needs to get out of the workshop for a while and rest.

- Your workshop should always have a fully stocked first aid kit.

In the next chapter, you will learn about the elements that you must consider when designing your knife to ensure that it is functional. We'll be looking at the different parts of the knife, along with the most common designs you'll find in use today.

CHAPTER TWO

DESIGNING YOUR KNIFE

Before we move into talking about the tools that we'll be using, we're going to first talk more about planning and designing your knife. This is a step that you can take before purchasing a single piece of equipment (apart from some graph paper and a pencil, but do those really count?). This is important because it means you can start bladesmithing before you spend any money on it. A proper design is the blueprint for your project, so it serves as the foundation for your bladesmithing work to come.

There is no real rule as to what makes for a good or a bad knife. One person may think that a spear point knife is a waste of metal, whereas another may adore it. A needle point knife might be hard to make, and thus sell for more money, but it may also have fewer customers interested in it due to its odd edge. So, to say that one design of knife is better than another is a pointless argument.

What is important is to understand the basics of the knife. If you get into bladesmithing and just start trying to make a knife without any consideration, you can still do it; however, the chances of you messing something up—such as not leaving enough space for a decent handle—becomes vastly increased. By learning your knife anatomy, sketching out, and designing your knife beforehand, you can ensure that you'll always make a great blade.

In this chapter, we'll first look at the basic anatomy of the knife, understanding that such may vary widely between different styles, but that all knives share the same basic core. From there, we'll move into the different types of knives commonly available these days to give you an idea of what type you are most interested in making. Beyond that, it'll be up to you to sketch out your design to ensure it meets all your needs.

BLADESMITHING

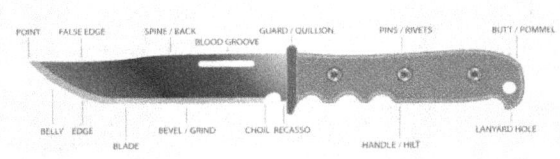

The Anatomy of the Knife

We'll be working from the front of the blade—what we call the tip—then work our way backwards, discussing the various parts of the blade, what their purposes are, and what they are called. This lesson will be crucial going forward, as we'll be looking at various designs and need to understand these terms if we hope to have any kind of understanding in terms of their differences.

Point: The point of the knife is the very last piece of the blade. It can also be defined as the part of the blade where the edge of the knife meets the spine. If you were to stab a pillow with your knife, then the point would be the first part of the blade to stab into the pillow, followed by the tip.

Tip: Tip and point could appear interchangeable at first glance of the word, but they actually refer to two different parts of the blade. The point is the sharp end of the blade where the spine and the edge meet. You could call this a tip and be correct, but that would only be if you weren't speaking in knife anatomy. When we speak about the *tip* of the blade in bladesmithing, we are referring to the front of the knife—usually about one-third of the blade as a whole. This would be where the blade goes from being thick and narrows towards the point.

Belly: The belly of the knife is nestled in nice and close to the point and the tip, though it refers to a specific part of the blade's edge. The belly of the knife is the part of the blade that curves as it gets closer to the point. This is, therefore, the part of the blade that is most important when it comes to cutting rather than stabbing.

Edge: The edge of the knife is the portion used for cutting. The edge of the knife runs the length of the blade from point to heel, typically. There are blades with much shorter edges, as well as double-edged blades with two edges rather than a spine.

Spine: The spine of the knife is the side opposite the edge. The spine is normally dull and flat, so you can use a finger or hand on it to get more force when cutting through tough material.

Bevel: The bevel is the part of the knife that is ground to create the edge. When you look at the edge of a blade, notice how it is thinner than the rest of the knife's cheek. This is due to the process of grinding it down to create the edge in the first place.

Cheek: The cheek of the knife is sometimes called the face. If you set your knife down on its side and look at it, the flat side of the blade is the cheek. This is the largest part of the blade, as it includes the entire blade. Typically, the bevel is not included as part of the cheek when describing blades.

Heel: The heel of the knife is a term much like the tip in that it refers to a specific section of the blade that can be larger or smaller, depending on the style. Whereas the tip refers to the upper section of the blade, the heel refers to the lower section of the blade. This section is roughly the last third.

Choil: If you look closely at the edge of the knife and trace your way down toward the heel, you'll often find the choil. It's a little notch in the blade at the end of the edge, but before the plunge line or the ricasso. The purpose of the choil is to make it easier for the owner of the knife to know where to stop sharpening the blade.

Plunge Line: The plunge line can be confused easily for part of the choil. The choil could be considered the part of the edge that then transitions into a notch and the notch itself, and the way it transitions into the ricasso can be considered the plunge line. If there is no sharpening choil, and therefore no notch, then the plunge line is the section of the blade where the edge ends and the ricasso starts.

Ricasso: This is the name given for the very last part of the blade before moving into the handle. The ricasso is the part of the blade, just above the quillon, which is unsharpened.

Quillon: This is most commonly referred to as the guard or the crossguard, but the proper term is quillon. It separates the blade from the hilt and helps protect your fingers from accidentally slipping from the hilt to the blade.

Handle: The handle is the term for the part of the blade that you hold. It can be made out of wood, plastic, horn, or any other material that you feel like working with, but it all serves the same purpose. The metal of the blade extends into a thin metal handle, which could be used just as it is, but it is always best when it is given a proper treatment, as this typically creates a much more reliable grip.

Tang: The tang of the blade is not normally seen once the blade has been completed, as it is typically hidden inside the handle, but it is the part of the blade after the ricasso that turns into a long, narrow shaft. The handle of the knife is put on over the tang.

Rivets/Pins: These are simply pins that hold both sides of the handle in place around the blade.

Lanyard Hole: Not every blade has a lanyard hole, but they are quite common. They make it easy to hang your blade off a hook on your workshop wall, or you can attach a lanyard to keep it close at hand when out and about in the wild.

Pommel/Butt: The pommel is the very bottom of the handle. If you stood the blade upright—so the point was facing the sky—then the pommel would be the part that the blade is standing on. Most pommels are curved to create a gentle surface without any sharp edges, but there are designs that make use of the pommel as a weapon for blunt trauma, thus making the knife an even deadlier device in those situations.

Popular Knife Designs

There are a lot of different designs for knives already being used these days. We will be looking at the most popular of those designs to see if there are any that standout to you as something you would like to make.

Remember that you don't need to follow along with any of these styles; just because they are the most popular doesn't mean they are the only options available. You can choose to create a knife unlike anything anyone has ever seen before, if that is what you want to do. Consider this section of the book as fuel for brainstorming; not a guide on what to make.

Bowie: Bowie knives were first invented as a fighting weapon by James Bowie in 1830, and they have continued to be extremely popular ever since. Since their invention, bowie knives have evolved quite a bit, and there are several designs that are used regularly. These blades tend to have large cheeks, an extended belly with an edge on both sides of the blade, and a quillon. However, the term "bowie knife" has grown so popular that many people will simply refer to any knife with a sheath, quillon, and clip point shape as a "bowie knife."

What is a clip point? Good question—let's look at the next blade design for the answer to that.

Clip: A clip point knife is extremely common, and the feature, as mentioned, is often taken as a sign that the knife is a bowie knife. This isn't necessarily true. A bowie knife may have a clip design, but a clip knife doesn't need to be a bowie knife. A bowie knife most often has a quillon to prevent your fingers from slipping onto the blade, but a clip knife doesn't need this feature. The key feature for a clip knife is that the blade is clipped back. This results in a much larger belly. A clip knife will have a shorter spine as a result of the larger belly, but the spine-side of the blade may or may not have an edge, depending on the design.

Drop Point: A drop point knife looks similar to a clip knife, only without such a pronounced shape. The drop point knife has a convex bend near the tip, and, typically, the edge of the blade takes up pretty much all of the blade, resulting in a cheek that is miniscule to the point of nearly being absent. This convex bend reduces the blade's ability to pierce; however, overall, the knife is easier to balance and keep stable with this design when compared to the clip knife. If you are looking for a piercing blade, then you won't want a drop point knife. Despite that downside, they do make good cutting tools, and you'll find that many of the pocket knives you see on sale will use this style of blade because it is quite versatile.

Hawkbill: This is a great style of knife if you are into camping. Its shape is perfect for shaping wood, cutting ropes, and even as a makeshift can opener if you forgot to pack one. This blade gets its name from how its shape distinctly resembles the beak of a hawk, thus "hawkbill." The spine of the knife goes from being straight to having a pronounced belly that curves downwards. The edge of the blade also extends downwards quite a bit further beneath the cheek in order to come to a talon-like point. A hawkbill blade doesn't typically have a quillon, nor a plunge line. The handles of these blades often have finger grooves worked into them to ensure the wielder doesn't lose their grip.

Needle Point: This blade looks almost like a mini-sword. Needle point blades are designed to be symmetrical, thus it would replace the spine with another edge. A needle point blade is pretty much useless as a cutting weapon; it simply doesn't have the strength necessary to cut because of how the tip tapers into the point. However, a needle point blade isn't for cutting; it's for stabbing. The design makes this an incredibly strong weapon when used to pierce, leading the needle point blade to be the most common design for attacking daggers (parrying daggers tend to focus less on piercing and more on defence, so they may or may not use a needle point). It is common to find needle point blades with jagged edges in the middle of the edge. Beyond simply looking cool, these edges make the needle point blade incredibly deadly because they rip and tear at the innards of whatever the blade stabs.

Normal: Even within the realm of "normal" blades, there is a lot of variation. However, for a quick and easy way to wrap your head around these blades, picture a typical chef's knife. The spine of the blade is flat for most of the blade—only arcing downward at the tip to make the point. The edge of the blade curves upward at the tip. The spine of the blade is mostly flat, transitioning into the handle without a quillon. The shape of the spine is important because these blades are explicitly for cutting rather than stabbing, thus the reason they are so often the design of choice in kitchens.

This design also allows you to use your free hand to push down on the blade to add pressure when cutting through tough materials. These blades are usually much heavier than the others we've looked at. This is due to the shape and size of both the cheek and the ricasso. Their size helps with cutting, but it makes the blade heavier, thus far less versatile. These blades are best left in the kitchen rather than taken along camping or hiking.

Pen Blade: This style of blade is named after how it is a much thinner blade compared to the others that we are looking at. Pen blades are commonly used for pocket knives, especially those that collapse into the handle. When looking at the spine of a pen blade, you may think that it is straight all the way through to the point, but there is actually a very small curve beginning at the tip. These blades are almost always only edged on one side, though it is common to find that the edge has a curved middle section reminiscent of the clipped blade, looking as if someone cut a semi-circle out of it. Although not great for cutting, these blades are good at piercing, and they are especially great for tasks like opening cans.

Sheepsfoot: These small blades are quite odd looking, as they don't have a point. The edge of a sheepsfoot blade is commonly straight. The spine is often quite straight, but it suddenly veers almost straight down to meet the edge at the tip. This results in a point that isn't so much a point as it is a wall. They're designed so that you can apply pressure and maintain control of the blade with a finger extended from the hilt over the spine. They're good for cutting and chopping, though they have absolutely no piercing ability whatsoever. Yet, this can still be a positive, since it makes it much harder to injure yourself or others accidentally with the blade, should it slip.

Spear Point: These blades are symmetrical, like the needle point blade. The belly of a spear point blade begins at the same spot, whether you are discussing the edge of the spine. Also worth noting is how, just because the shape of a spear point blade is symmetrical, that doesn't mean the blade itself is. You can absolutely purchase spear point blades that have two edges, but it is also incredibly common to come across spear point blades that only have a single edge. Like many of the blades we've looked at, spear point knives are better at cutting than they are at stabbing. However, this particular design is more aerodynamic than the others we've looked at, which is why many throwing knives use this design.

Spey: Spey point blades are included in many Swiss army knives. They're terrible at stabbing and piercing thanks to their shape. Rather than coming together to form a sharp point, these blades typically only have a slightly curved belly, which results in a rather dull point. They were designed for practical use rather than as a weapon. The name of these blades can actually tell us what they were used for, and they would make Bob Barker proud. The design of the spey point blade made it ideal for neutering animals on the farm, so it was much more common for ranchers and farmers to have spey point blades around. Along with the neutering, this design is also incredibly efficient when it comes to skinning animals. Although other livestock knives, such as the saca tripas, are effective fighting weapons, the spey point blade simply is not.

Tanto: This blade took its inspiration from Japanese swords. Of all the blades we've looked at, this is the one that would scare me the most to see, should I ever be facing someone armed with it. This is thanks to the strength this blade offers, which makes sense considering it is a fighting weapon, and even its name points toward this reality. Cheaper samurai swords were known to break in the middle of combat, yet the tip of a samurai sword could still be a fierce stabbing weapon, able to pierce through armor and win the day. The tanto blade is designed after this bloody history. It looks almost like a reverse sheepsfoot blade, with the point up by the spine rather than down by the blade. Yet, the belly of the blade is still mostly flat. Being flat, the belly does not offer any cutting ability, so it is useless to even try. However, its design makes the tip of the blade incredibly strong, earning the tanto blade the distinction of being the deadliest stabbing weapon on this list. The point is so strong that these blades can easily punch through the door of a car—a terrifying thought if ever there was one.

Trailing Point: The trailing point blade is an incredibly cool looking design, and one that is shockingly effective at cutting. These blades can be deadly fighting weapons, though they are also fantastic for skinning animals, just like the spey blade. There is almost no cheek to the trailing point blade. The handle moves into a choil, whereas the rest of the blade is edge. The spine of the blade has a slight curve too, like a valley, so it starts curving downwards after the hilt but then curves upwards toward the point. The edge of the blade extends from just after the choil through to the point, and it takes up almost all of the cheek. In fact, the tip of a trailing point blade is pretty much all edge with no spine to speak of. Although these blades can be found with straight handles, it isn't uncommon to see handles that arc upwards for an inch, and then turn downwards to create a stronger pommel that can be used for striking.

Wharncliffe: The wharncliffe blade is a fascinating one with a unique design that makes it a great blade to finish off on. These blades have large ricassos that jut out slightly beyond the edge in a manner that serves almost like a quillon. The edge runs from nearly all of the blade through to the point, though it still stays straight. The edge itself takes up almost all of the blade, leaving only the smallest of cheeks. The spine begins to curve toward the point, pretty much the second the handle stops. This curve begins gradually, but it gets a little more extreme toward the point. Although this blade is mostly edge, the spine's angle actually makes the point less dangerous, as the belly of the blade is designed not to increase cutting power, but rather to decrease the chance of accidental stabbings. It might come as no surprise then, that these blades were primarily used by sailors who needed a blade that wasn't going to kill them, should the waves of the sea toss them around the deck.

Chapter Summary

- Each part of a knife has a specific name. Many of these words are used interchangeably by the general population, so it can be difficult to tell if someone is referring to the right part of the blade when using a term like "tip." Bladesmiths will always strive to list the anatomy of the knife accurately.

- The point of the knife is the very end where the spine and the edge come together to form a point. This is the part of the knife that enters first when stabbing something.

- The tip of the blade refers to the front of the knife, about the one-third of the blade from the point toward the hilt.

- The belly of the knife is the part of the blade that curves toward the point. It is this part of the blade that is used for chopping vegetables and the like.

- The edge of a blade is the portion used for cutting, running from the point to heel.

- Opposite the edge of a blade is the spine. This is the part of the blade on the top that is not sharpened. Some blades are double-edged, and thus don't have a proper spine.

- The part of the blade that we grind to make nice and sharp is referred to as the bevel.

- The flat part of a blade is called the cheek. When we grind a blade, what we are doing is turning the cheek into part of the bevel.

- The heel is the lower part of the blade. It is the reverse of the tip.

- Some blades have a choil, which is a part at the end of the edge to let you know where you stop sharpening the edge.

- The plunge line is a small notice in the blade at the end of the edge. It transitions the blade from the edge to the ricasso.

- The ricasso is the last part of the blade before the quillon and the handle. The ricasso is to the heel similar to how the point is to the tip.

- A quillon is also known as a cross-guard. This part of the blade is more pronounced on swords than on knives, but its purpose is to prevent your hands from slipping onto the blade.

- The part of the blade that you hold is called the handle.

- The tang of a blade is the part that the handle is attached to.

- Many blades show rivets or pins, which hold the handle in place.

- Some blades have a lanyard hole—a simple hole that is used to attach your knife to a keychain or the like.

- The pommel or butt of the handle is the very bottom of the knife.

- The bowie knife is a type of fighting weapon with a large cheek and extended belly. They may be single or double-sided and typically have quillon.

- The clip blade is one that had a part of its blade chipped back to create a larger belly.

- The drop point blade has a convex bend near the tip and an edge that takes up almost the entire cheek.

- The hawkbill knife has a pronounced, downward curving spine that gives it the sharpness of a hawk's bill or talon.

- The needle point knife is like a tiny sword. It doesn't have much of an edge, but it comes to an extremely powerful point that makes it great for stabbing.

- The normal blade has a straight spine with a slight downward arc at the end to create a sharp point. These blades make for great kitchen knives because their design allows the user to apply a lot of pressure for cutting.

- The pen blade is thin, extremely common, and found on most pocket knives. They have a very small curve with a single edge.

- The sheepsfoot blade doesn't really have a point; rather, the edge remains straight and the spine often takes a sharp turn down to meet it. There isn't really a point, though more a flat line. It can't stab with any real strength, but it can cut with the best of them.

- The spear point blade is symmetrical with one or two edges.

- The spey point blade is found commonly on Swiss army knives. These knives were used to neuter or skin animals.

- The tanto blade is a type of Japanese stabbing weapon that is so powerful, it can stab through a car door.

- The trailing point blade is amazing at cutting, due to how the blade curves.

- The wharncliffe blade has a large ricasso with a straight edge that takes up most of the blade. This blade is more edge than knife.

In the next chapter, you will learn about the different tools that you need to start making your own blades and knives. This can be an extremely expensive skill to begin, so I have broken down the tools required into three sections: beginner, intermediate, and expert. These categories will increase difficulty and price accordingly.

CHAPTER THREE

TOOLS NEEDED

Getting into knifemaking can be a difficult process for beginners who aren't sure where to start. Finding the space, figuring out what tools you need, and the types of blades you want to make are all considerations that can make it difficult to start making knives as a beginner. It takes time and effort, and the beginner is forced to wade through all sorts of different opinions. Some people say you should invest in a forge out the gate, whereas others say the forge is one of the last things you will need. Some recommend a disc sander when others will suggest that you can get away with using sandpaper.

So, just which of these tips exactly are right?

In a lot of ways, both of them are. Knifemaking is a skill, but it is also a creative pursuit. You have to follow some key rules when you're making your blades, since this is a more concrete skill than, say, writing a novel, but you are encouraged to mess around and try new things, styles, and approaches. If what you really want to do is forging, then that's the right path for you.

However, that will be an expensive path, and many beginners don't want to drop that kind of money on a skill they are only just starting to pick up. In order to make this the easiest possible experience for you, this chapter will be split into three parts, which can be thought of as cheap, costly, and expensive; or it could be thought of as beginner, intermediate, expert.

The first category is for the beginner. You can expect to pay somewhere between $500 and $1000 for these tools, though there are corners you can cut such, such as borrowing tools or waiting for sales. The next category, intermediate, is made up of the tools from the beginner's level, plus a few others that could end up costing you nearly $5000. The final section, the expert, is for those who are looking to forge their own knives, which could cost you about $3000. Remember that each of these categories builds on the next, so if you bought everything and paid the maximum price, then you would be looking at roughly $9000 worth of tools.

That's a heck of a lot of money. So, let's break it down category by category.

The Beginner's Gear

This level of blademaking gear won't break your bank when buying, but it will be noticeably limiting. You simply just won't be able to make every type of knife. You will also have a much harder time customizing and making unique knives, unlike those we've seen in the world, especially if that is what you're into. However, what you will get out of it is enough gear to get the basics down and start making some simple blades. Despite being simple, these blades can still be sold for a decent amount of money and earn you some nice income if that's what you're after.

The biggest downside with this level is that, unless you are making only small knives, then you are probably going to have to send them out to get them heat treated. When we use a knife, it is typically pretty hard. However, when we are making a knife, it doesn't start out that way. The heating process—which we will look at more in chapter five—strengthens and hardens the metal. Small knives may be able to be heat treated with a simple flame, but most blades will require heat treatment with a forge, and that's not until the expert level. So, before we get there, let's see which tools we can get our hands on at this level.

The first thing you'll want to get as a beginner is a workbench. We spoke about how important tables and surfaces were in the first chapter, but it is still worth reiterating. You will be working with blades and tools that could damage a table, so get a workbench that is designed to withstand that kind of pressure. Also, you may want to consider investing in a comfortable stool, since you'll be finding yourself at the workbench quite often.

The first tool that you'll be looking to get is a five-inch angle grinder. This is a power tool with a circular disk on the end that spins really fast, so you can grind out shapes for your knives and cut away metal. A grinder is only as good as its discs, though. It's like a drill that way; you could have a drill, but if you don't have any drill bits, then there would be no point. At this stage, you won't need to drop a lot of money on discs for your angle grinder—just pick up some one-millimeter discs for your grinder, and you'll be good to go.

Next up is a set of files. This is one of those tools that you'll want to invest in purchasing high quality rather than cheaply. You'll use these files to perfect your knives' bevels, and this can take quite a bit of time. Similarly, you will want to purchase several kinds of sandpaper. We'll be using sandpaper in place of a grinder to begin with, though we will want to work our way up in grit count. Start with sandpaper that has a grit count of 180, and then move up to 320 and 600 grit count. As you work through your blades, you'll begin with the lowest and move up from there, rather than perhaps starting with 180 as a beginner and moving onto 320 as an intermediate. The sandpaper is just for the beginner—intermediate bladesmiths won't need to worry about getting any.

The angle grinder could very well be the most expensive tool you purchase—that is, if a small drill press doesn't run you more. You want to make sure that you make straight holes, and so you should *absolutely* invest in a drill press. Before the metal is hardened, it is easy to drill through, so don't worry about breaking your tools on it. However, if you really want to make sure that you get everything working properly, then you should also get a bench for your workbench. This will ensure that your blades don't move when working on them, which is especially important when working with an angle grinder.

At this stage, you're only missing one thing: metal. We'll be using steel in a moment, but we won't be worrying about raw metal until after our first couple of knives. For those, beginners should use pre-cut blanks. A blade blank is a piece of metal that has already been shaped and cleaned. You'll still need to grind them down with your sandpaper, use heat to harden them, and then finish them. However, the first two steps—the designing of the blade and the shaping of the metal, have already been done. Blade blanks are a great way to get a jump start on making your first few blades, and they'll be great for training those later stages of completing a knife. However, they have one more great feature, which is that they allow you to get a hands-on look at what a proper knife is supposed to look like. That would include how the blade is shaped and the tang looks. This will help you in designing your own blades later on.

Speaking of later on, you won't want to use blade blanks all the time, since they can only take you so far. If you want to improve your skills and move beyond being a beginner, then you will need to learn how to cut and shape metal yourself. For that, you'll need the metal. We'll use steel, though we want to ensure we use one that is easy to work with. That means we'll use three-millimeter 1075 steel. It is a good idea to use this number for your blade blanks, too, as it is easy to harden, so you can probably manage to do so with a torch rather than requiring a forge. However, what do those numbers mean? That number is the SAA steel grade used for carbon and alloy steels. The first number tells you what the main alloying element is; the second digit is the top grade element; and the final two digits let you know how much carbon is in the metal, though these two numbers represent a percentage to the hundredths. If we look up a SAE guide, we would find that 1075 steel is a plain carbon steel that contains 0.75 wt% carbon.

With these tools, you have what you need to cut and prepare your own knives. You'll have to use the sandpaper instead of a grinder, and you'll need to use a torch or send your knives away to harden them. Sandpaper is a perfectly fine and acceptable way to grind your blades, though it is incredibly time consuming. The problem of hardening your knives can prove to be tricky, but at this level, you simply shouldn't be investing in a forge yet anyway.

Intermediate Gear

Remember when going up to a new level of gear that you won't be getting rid of what you bought previously. This next level is an addition on top of what you have already purchased. However, with that said, it does replace some of your earlier gear. If you are someone who is looking to purchase all the necessary gear out the gate, you may want to consider purchasing the two tools that make up this section early because they would replace your angle grinder and sandpaper you were using to finish your blades.

As mentioned, it takes a lot of time and energy to sand down your blades by hand. It is absolutely and undeniably a method that works, but it will easily eat up dozens upon dozens of hours over the course of a working year. Frankly, that is time better spent doing other work. So, one of the tools that intermediate bladesmiths should look to purchase is a disc sander. A disc sander is a tool that you would mount onto a workbench in the same way you would mount a vice. The main part of the sander—the disc—typically juts out over the edge of the workbench, so scraps can fall to the ground rather than onto the desk. The disc spins quickly, and all you need to do is press the part of the blade you are looking to sand into it, and it's spinning will do all the work. This function makes sanding incredibly easy. Considering that we're looking at bladesmithing, you shouldn't need a disc sander larger than nine inches.

The other tool that makes up the intermediate kit is a belt grinder with a variable speed. Since, at this point, we are committing ourselves to spending a large amount of money on our bladesmithing skills, it is absolutely worth purchasing the highest quality belt grinder that money can buy. Otherwise, you should maybe at least look toward the highest quality belt grinder that doesn't break your bank. There is a generalized rule that professional bladesmiths follow: a high quality grinder can hide the effects of a poor forging job, but a high quality forge can't make up for the damage a low quality grinder will cause. It is always best to purchase high quality. It could be more expensive, but you'll see that it is worth it by the quality of the blades you create. You'll also need to replace a high quality grinder far less often than a low quality one. In fact, if you are lucky, then you will only need to replace it once or twice over the course of a lifetime. Those who are incredibly lucky won't even have to do this. However, you will need to replace your sanding belts. These will end up costing you more than the grinder eventually, and they can't be made to last indefinitely. On the other hand, there is one more piece of gear—albeit more an accessory than gear—that can increase the longevity of your belts and improve both your grinder and your sander.

That accessory is a variable frequency drive. This is an electronic device used to control the speed of the motor in a electro-mechanical system, such as a variable speed belt grinder or a disc sander. A variable frequency drive would allow you to fine tune the speed of your tools, which has two major benefits: the first is that it allows you to increase the kinds of material that you can use when creating your knives. Different materials need different speeds because too fast can damage some, but too slow will be useless for others. By having full control over the speed, you open your workshop up for more business. The other cool thing that a variable frequency drive does is save you money. You will be wearing your way through your belts. The more knives you make, the more belts you'll find yourself burning through. However, a variable frequency drive actually helps you extend the lifespan of your belts by slowing them down a bit, wearing them away less with each blade.

A variable frequency drive is pretty expensive. Buying one for your sander and one for your grinder could easily cost you over $1000. However, if you are careful when picking out your gear, you can actually get away with only purchasing one. When buying your grinder and sander, keep an eye on the type of motors they use. If you can purchase gear that uses a three-phase motor, then you can actually use one variable frequency drive on both pieces of equipment. This is a great way to increase your productivity, save money, and maximize your equipment to work at the highest possible level of quality.

This kit will take you a long way. It opens you up to work with more than just 1075 steel, but it isn't perfect yet. You'll still need to either torch heat your blades or send them away to be heated by a professional with a forge. This isn't the worst possible downside. After all, it can be a great way to meet someone else with an interest in bladesmithing (or just blacksmithing). Someone who already has a forge likely has plenty of hands-on experience and knowledge, which you could benefit from. This could make for a great friendship, though you need to remember to ask lots of questions!

But in the end, you will find that sending your blades out takes more time than it is worth. You gotta either drive them out or ship them out to get heated. This means travel time, waiting for them to be slotted into a busy schedule, and then more travel time to get them back. Just like sanding by hand, there is a more efficient way to heat your blades. However, it will cost you a pretty penny.

Expert Gear: Forging

The first thing you need to understand about forging is that you don't actually ever need to get a forge. Yes, it will save you time on heat treating your blades, but the tools you already have should be enough to make your blades. You have to get them heat treated elsewhere, but the creation of those blades—from design to shaping, sharpening, and finishing—all of that is possible to achieve with the gear that we've already looked at. A forge isn't a necessary purchase, but it is a really cool one, and many people find that they eventually want to get one.

At this level, you should have all the gear from the beginner level. You don't necessarily need to purchase the gear from the intermediate level, but I do recommend it. Purchasing all of this gear is extremely expensive, so don't worry if you can't purchase it all at once. For this level, we'll look at the gear you'll need. However, there will be a little twist at the end that can make getting a forge both a profitable and enticing idea.

So right out the gate, you'll need the forge. A forge is an object that traps heat inside of itself. In the past, they were constructed out of stone and brick. They still can be, but we also have many other options available to us in this modern age. They were commonly heated with coals in the past, but modern forges are more prone to using gas. You can expect to pay between $500-$1000 for a forge. It will require a space for installation—one with temperature control and ventilation as described in chapter one. Investing in a forge is a one-time purchase, but heating and keeping the forge working will continue to cost you money, as you'll need to refill your gas when it is running low.

You may be shocked to find out that the next item, an anvil, costs even *more* than a forge. This is because an anvil is priced based on the pound, and they are made out of heat-treated steel. They can easily weigh several hundred pounds, with larger ones pushing upwards of a thousand pounds. The forge is used to heat your metal.

It is then moved over to the anvil, where it is beat into shape to remove any impurities from the metal.

To beat the hot metal, you must have a hammer, and not just any hammer will do. You will want a hammer that is designed for use with metalwork and forging. Not only that, but you will need to get your hands on a couple of them because not every hammer will be right for every blade; it just doesn't work like that. You will want to pick the right hammer for the right job, and in order to do that, you'll need to have a couple hammers from which you can select.

Similarly, you'll want to get some tongs. These are simply to remove the red hot metal from the forge without burning all the skin off your hands. You wouldn't even want to reach into the forge with gloves on because it is simply too hot. A pair of metal tongs will do the trick for you, though you must still wear gloves, since they absorb heat themselves. I recommend purchasing two or three pairs of tongs out the gate because you'll never know when you'll need the extra.

The final piece of the expert's kit is a flypress. This is a type of screw press that uses a flywheel or fly weights to help drive the screw through the material. This tool is used to punch holes in a sheet of metal in one quick jab. If you were to cut the hole, then it would cause more damage around the edges—same with drilling. However, a flypress pushes the screw straight through in one quick punch. It would be the difference between using a hole-punch on a sheet and just stabbing a hole in it with a pencil. The latter could still get the job done, but it will be of noticeably lower quality.

So, although that is the gear that makes up the expert kit, there is still one more thing we need to talk about in this kit before we move onto the extra gear you may want to consider purchasing down the road. Did you notice how we have tongs and hammers both listed under this section? You will need to purchase these to begin with because you can't beat metal without a hammer. However, you shouldn't go out and purchase every single hammer or set of tongs that you see. Begin by first purchasing two or three of each. Having these will let you get started working your forge. Then, you can benefit in the coolest manner. That hammer you're using? Those tongs you've just purchased? Both of them are made through blacksmithing. They have been designed, cut, forged, and finished, following an incredibly similar workflow to the blades you've been making. Clever readers will already see where this is going.

With a little bit of extra training, you can take that forge you bought for bladesmithing and use it to make your own tools. If you need a larger hammer, you don't necessarily need to go buy one; you can use the equipment you already have to make your own. Purchasing a forge is pretty sweet because, although it is

BLADESMITHING

quite an investment, it can actually save you money on forging down the road. Plus you can always sell the blades and tools you make with it, and—double plus—you can rent it out to other bladesmiths to do their heat treating for them. This level might be the most expensive but it definitely opens up the most doors going forward.

You could absolutely stop purchasing gear right here if you wanted to. In fact, you could stick with just the beginner's gear if that was what you wanted. However, if you've made it to this level, then chances are good that you will be continuing to expand your workshop. We now turn our attention to some of the extra gear we can invest in to improve our productivity and capabilities beyond the level of expert.

Extra Gear or Alternatives

The gear in this section is in no way necessary for bladesmithing; however, it could prove to be incredibly useful. For example, cutting and shaping a hilt can be easier with some of these tools. Even shaping the blades themselves can be made easier with the information provided in this section. But easier doesn't necessarily mean they are required. For some people, the amount of money they save by foregoing these tools is worth the added time it takes to create their blades. Which of these is right for you will depend on what you are willing to spend, where you want to make things easier for yourself in terms of creating, and which way your interest in bladesmithing pulls you.

Let's start with the bandsaws. The more attractive bandsaw for many bladesmiths is the steel bandsaw. Easily costing you over a thousand, they are an expensive piece of equipment. However, being able to purchase large sheets of metal and cut it into workable sizes in a matter of minutes is hard to beat. But that price tag is definitely rough. You may also want to consider getting a wood bandsaw for working with hilts and the like. On the other hand, if you already have a steel bandsaw, then you won't need to purchase a wood bandsaw. A steel bandsaw can cut wood, but a wood bandsaw cannot cut metal.

What if you wanted to heat treat your knives yourself, but you weren't interested in purchasing a forge? Well, then you could purchase a heat treatment oven. Just like the oven in your kitchen, this type of oven traps heat inside, though its purpose isn't used for cooking. It gets hot enough to treat your blades, but it doesn't get them to the extreme heat that a forge needs to reach to work the metal. If beating the red hot steel with a hammer isn't appealing, then this is a good option; however, it is extremely expensive—nearly $3000 on the low end. Plus, you'll still need to purchase tongs to use with it.

An extremely expensive option—this one upwards of $7000—is to get your hands on an extremely strong hydraulic forging press. We're talking about 25 tons here. A weaker hydraulic forging press will cost you a lot less money, but it will also keep you from reaching a further level until you upgrade. This is a way of shaping metal that basically replaces the need for beating the metal against an anvil with a hammer. This tool uses hydraulic pressure to push two slabs of metal against each other and flatten out the object inside. In other words, it beats the metal for you. If you have issues with your upper body strength, then a hydraulic forging press could be an appealing option to achieve the same effect without hurting yourself.

Another option is to purchase a strong power hammer. This is a forging hammer that is controlled mechanically rather than by swinging it yourself. For bladesmithing, you'll want a powerful one with about 21 kilograms of weight behind it. The mechanics lift up the hammer, then drop it down onto the surface built into the tool. If you need something hammered, then you would use the buttons to lift up the hammer, stick your object beneath it, then hit it as much as you need to. The machine does the hard work for you, but for this convenience, you can expect to pay as much as $10000.

When you are working with wood, you would use a planer to make the wood less thick. With metal, we would use a steel rolling mill. Metal is fed into this piece of equipment and rolls inside of the machine to thin it out. If you have ever rolled dough in the kitchen with a rolling pin, then you've seen this process in action, albeit in a tastier form. The same idea applies here, only you probably shouldn't eat the steel once it comes out of the oven.

One of the key components of bladesmithing is the hardening of the metal blade. But just how hard do you want it to be, and how do you know it's hit the right level? To answer these questions, we would use the Rockwell hardness test, which can be performed with a Rockwell hardness tester. There are a lot of Rockwell hardness testers on the market that you can use to test your blades. They will give you back the hardness by measuring the depth of indentation it causes. This rating will be determined by the Rockwell scale—one of the most popular and widely used hardness scales in use today.

Finally, if you are into making sheaths for your blades (which we will look at more in chapter seven) then you may want to consider getting a leather sewing machine. Just as it sounds, it is just a sewing machine that is designed to be used with harder materials, such as leather. You can absolutely do this sewing by hand, but a leather stitcher can save you a lot of time if you are making lots of sheaths.

Together, these make up the most commonly seen extra tools in a bladesmith's shop. They don't offer you anything you can't do yourself with expert level gear, but they can make doing those things much easier. That said, you're looking at a price tag of about $25000 or more. That's not counting the gear from the beginner, intermediate, or expert levels; only the extra gear showcased in this section. That type of money is hard to take lightly, so if you are looking to expand, your best bet will be to be slow and steady with an eye that would continue looking out for deals.

BLADESMITHING

Chapter Summary

- Getting into bladesmithing can be incredibly expensive. To make it easier, you can begin by purchasing the beginner's gear, and then moving up to intermediate gear, and finally expert gear.

- A beginner's gear kit will cost between $500 and $1000, but it'll give you everything you need to work on your blades except for heat treating. You can always send your blades away to get heat treated or simply use the makeshift forge described in chapter five.

- The first piece of gear that you need to get is a workbench. It is here that everything would happen.

- A five-inch angle grinder with some discs makes for an easy way for beginners to start grinding their blades to create their edges.

- A set of files is important to work on the bevel of the blade. Sandpaper ranging from 180 grit to 600 grit is also important for this reason.

- A small drill press is a great investment, which would allow you to make quick and easy holes.

BLADESMITHING

- You should also get a vice grip and mount it onto your workbench, so you can keep your blade steady while working on it.

- You'll need metal to begin with. A 1075 steel is a good starting point, and you can purchase some blade blanks to get a quick start on your first blades.

- Steel is graded according to the SAE guidelines. These guidelines assign a four-digit number to the metal. The first number tells you what the main alloying element is; the second digit refers to the top grade element; and the final two numbers tell you the percentage of carbon in the metal to the hundredths.

- The intermediate gear kit adds two new tools: a belt grinder with variable speed and a disc sander.

- The disc sander will make it easier to sand your blades, since you won't need to do it by hand anymore.

- A belt grinder greatly improves the speed and quality of your grind, so you can get the best edge possible on your blade.

- For expert gear, we would move into forging. Forges can be powered by gas or coal.

- Most of the gear that you purchase as an expert would be to improve your ability to forge. Hammers, tongs, an anvil—all of these are directly related to the forge itself.

- Another useful tool that experts invest in is a flypress. A flypress would replace your drill. This doesn't drill into the metal; rather, it quickly and efficiently presses a screw through it to create a better hole.

- There is a lot more gear that you may consider purchasing as you pass the level of expert, but it would start to get incredibly expensive at this point, and it really isn't necessary to purchase anymore gear to become an expert bladesmith.

In the next chapter, you will learn how to grind a knife. Grinding your knife is how you would create a sharp cutting edge, and it is an extremely important part of making your own blades. However, it is incredibly easy to ruin your knife at this stage; that's why we'll look at an easy-to-follow guide that'll make the process much smoother and less prone to error.

CHAPTER FOUR

GRINDING

When it comes to grinding a blade, there are many options available for how you can go about doing this. There are all sorts of different grinds that we could use—from hollow grinds to compound grinds, onto high flat grinds and asymmetrical grinds. Don't worry if these don't make sense yet, we'll be looking at them shortly.

The secret to grinding a knife properly is to be patient. We need a good grind if we want the blade to be sharp, but if we push too hard at this stage and rush it, then we will end up with some wasted blanks. It is important that we plan out our grind well before we begin. When we rush to get started, we're prone to making mistakes.

There are several ways you can grind blades, but we'll be looking at a specific technique in this chapter.

We'll be using a belt sander instead of an angle grinder, though you can absolutely use an angle grinder by adapting this approach just slightly. We will also be using a machining dye to help us find the center of our blade. If you don't have machining dye, don't worry—you could use a permanent marker to achieve the same effect.

We'll start by looking at how you grind your blade in a way that will ensure you end up with a great bevel and edge. To achieve this, we will be using the grinder a total of three times. We'll begin with shaping out metal, then grinding our edge, and then with a second grinding pass to finish the blade. This will walk you through all of the steps involved in grinding, though you'll need to read chapter five before you get to the third and final part of the grinding process.

After this, we'll close out the chapter with a quick look at the more popular styles of grinds, like those we mentioned at the beginning of the chapter.

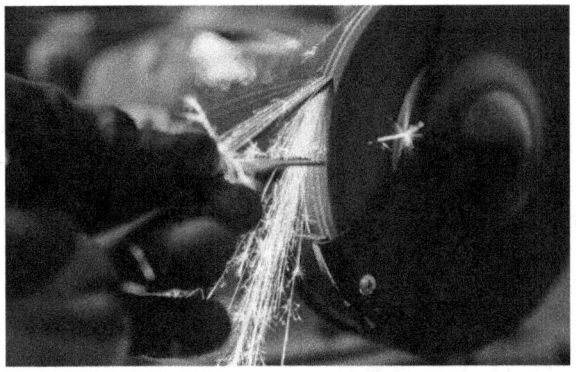

Step One: Grinding Out the Design

Before you can begin to worry about grinding the bevel, you must first come up with a design for your knife. In chapter two, we looked at the most common designs for modern day knives. There is quite a bit of range to these, and deciding which to begin with isn't always easy. If you can't decide for yourself, then go with a normal knife or something equally simple.

To design your knife, you will want a pencil and some paper, and all you need to do at this point is draw out your knife. However, for those just beginning and those who aren't so confident at drawing, this may be easier said than done. One way you can make this part of the process easier is to use a specialized software like Knifeprint. Knifeprint is a program made specifically for designing knife blades. You would use the editor in it to shape the blades and pick where the various elements are. It is free to use, so long as you save your projects publicly. Users that want to save their knives privately will need to purchase a lifetime or annual license. Since there are many public projects on the site, you can select the knife you want to work on easily and print out the design. We will be using this method; however, you can also use Knifeprint to purchase laser cut blanks in the design of your choice if you aren't interested in grinding out the shape yourself.

Once you have your knife design printed out, use some machining dye to stick it onto your raw steel. This would give you an outline for your knife that you will be using in conjunction with the grinder. Most designs will have rivet holes that must be drilled out as well, don't forget. Before we begin grinding, it can be a good idea to cut the shape further down in size, so that your raw metal is already a little knife-like. If you have the tools for cutting metal, this would be the time to use them. Otherwise, start up the grinder.

At this stage, all we are looking to do with the grinder is get the shape of your knife down. This means that we want to be grinding it while it is perfectly level. An angle at this point will start to form and edge, and we're not ready for that yet. As you grind the shape of the blade out, you should have a flat spine and bevel. There may be a curve to the blade, but that curve should not have an edge yet. The point should be a little sharp—

as it is made out of an angle—but it should have absolutely no cutting ability yet.

At the end of this stage, you will have a piece of metal that is clearly a knife, but we won't be ready to sharpen it up yet. First, we must take a brief break from our grinder to determine the placement of our center and, therefore, the edge.

Step Two: Finding the Center

This step is pretty easy. The goal is to find the center of your knife, but we don't mean the center of the cheek. Looking at the bevel of your knife where your edge will be, we want to mark out the center, so can we have a place from which to work.

The next step in this process will be grinding out the edge, and this will require us to be extremely careful. If we grind too far in on one side, then we'll pass the center and, ultimately, change the shape of the blade. Try grinding out an edge without marking the center and see how quickly it takes you to mess it up. It is extremely difficult because it requires you to eyeball it perfectly, and we're beginners; we're definitely not up to that level of skill yet. In fact, there aren't many experienced bladesmiths that can eyeball their grind accurately either. So, if we don't want to waste metal unnecessarily, we will need to have a guide in place.

Thankfully this is incredibly easy. Take your machining ink and coat the bevel. This ink will wash off well before we're done making our blades so don't worry if you think it looks bad. The reason we do this is so we can mark our center. If we tried to mark the center without first using ink or a permanent marker, then we would have to scratch the blade, and this isn't ideal. Applying ink to the blade lets us mark the ink instead. This keeps our center line clear without damaging the blade. It also will make the grinding even easier because we would use the ink as a guideline on how the bevel is shaping up.

Once the ink is on your blade, you must mark it. There are a few ways we can do this, but the most common strategy is to use a center scribe. This is a small piece of metal with a couple of protruding bolts. The bolts go against either side of the blade, so the sharpened pin between them can mark out the center of the edge. You can purchase a center scribe pretty much anywhere you can find blacksmithing equipment, or you can simply check Amazon to find dozens. Although this is a perfectly fine method, you can actually save money and be just as effective (if not more so) by using a drill bit. Set your knife flat on its side and place the drill bit onto the workbench. Then, take the end of the drill bit and run it along the bevel. This first pass will result in a line that is technically along the center, though it will favor one side more than the other. To make this even, simply flip the blade over to the other cheek and do another pass with the drill bit. Now, you should have a line that is truly centered, and your blade will be ready for the next grind.

Step Three: Grinding the Edge (Pre-Heat Treatment)

This is the last grind that we would make before we turn the blade over to be heated and quenched. It is this pass of the grind that will give us the edge we are after. In this section, we are going to assume that the edge we are looking for is a simple design that takes up most of the cheek, such as is common with knives of the clip point, drop point, or trailing point style. If you are looking to create a different style of blade, then you will need to adapt this step for your own purposes. Don't worry; that isn't very hard. Just pay attention to the key tips and tricks, and you'll have no problem altering this step for your own needs.

In the last step, we marked the center of the edge. In this step, we will be using that mark to make our edge nice and sharp. We would start by turning on our belt sander and getting it ready. The last time we used the belt sander, we made sure to hold our blade nice and level. This time around, we will be starting with a pretty steep angle. We only want to grind away a little bit of the blade at this point, but we want to make sure that the angle is nice and steep, so that it is clearly defined. At this point, you also don't want to cross over that middle line. Leave just enough space between your grind, so as to not grind any part of this line.

If you have done this properly, then the center line will still be flat, but the side of the blade will be noticeably missing some of the metal, making it look almost like a small hill. At this stage, you could flip sides and do the same grind on the other. Some people like to work each part of the grind on both sides before moving to the next; others prefer to complete one side of the blade before moving on. Regardless of the approach you take, the process for each side will be the same.

The next grind we want to make will be even steeper, traveling over much of the cheek of the blade. The first grind should only be small; simply getting us away from the center line, so we don't risk the integrity of our blade. This second grind begins from the end of the first and covers all of the cheek that will be part of the bevel. This requires us to press the blade against the grinder at a much smaller angle. If we had started this way, then hitting the center would have been incredibly easy; however, if you look closely with this second grind, you'll see that the center is nowhere near the grinder.

At this point, we have the equivalent of a rough draft of our bevel. But if we stopped here, it wouldn't really be a nice knife. The first angle of the cut is quite acute, whereas the second angle is much less so. The problem here is that it isn't really sharp yet. Nevertheless, if you take the knife away from the grinder and look at it, then you'll see that there is noticeably a lot of metal missing. The second grind shows us where the grind line will stop. This gives us a sense of what the final product will look like but we aren't finished.

It will take two or three more passes for the edge to be finished. Each of these passes is designed to remove the first grind we did. That acute angle we started with actually hurts us at this point rather than protects us. We needed it to begin with because, without it, we would have a much harder time setting our designed grind line with the second pass. However, now that the line is set, we want to remove it. This requires us to go over the blade a few times until the edge begins from our center marker and runs over the cheek of the blade to where our grind line stopped. What was once made up of two angles arcing away from the center line will thus become one singular angle, giving us our edge.

We must do this on both sides of the blade to have a real cutting device. Doing so will turn that center line into a sharp point that can cut with ease. Remember to be careful when grinding and take your time to ensure you don't cross the middle line; otherwise, you will have to start all over again.

If you have been successful with this grind, then the next step in creating your blade would be to heat and harden. We will look at that in the next chapter. But after the hardening of the blade, we need to work it once more, so let us turn our attention toward this final grinding session.

Step Four: Finishing the Blade (Post-Heat Treatment)

Since the last step, you've heated, cooled, and tempered your blade. This is a whole process that we'll be getting into in the next chapter. But after that, we want to finish the blade, and this means we have to get it back over to the sander yet again.

This time around, we'll be doing a series of passes over the blade, being extremely careful with our sander and working our way up in grit count. This same technique is used when sanding wood to create a smooth and seamless finished product.

We will no longer be grinding; instead, we'll be using our sander. This is what we would use to finish the grind and make it high quality. We would start with a sanding belt rated to 80 grit, which will help us remove the larger imperfections. As we continue in this fashion, each time we go up in grit count, what we are doing is removing smaller and smaller imperfections. So, we would first do a pass with a grit count of 80. We would follow this with a count of 120, then another of 200. After we hit 200, we would switch from sanding belts to what are called gator belts. These belts have large grit counts, but they aren't as tough as your regular sanding belts. This means that you won't run into the issue of the belt causing damage to the blade. Gator belts are rated a little differently, but for a blade, you'll want to get A65, A45, and A30 belts. These are equivalent to roughly 200 grit, 400 grit, and 600 grit respectively.

During this next pass, you won't need to worry about damaging the blade the same way you had to when grinding it because the sander doesn't remove metal in the same way. Plus, it has been hardened, making it much harder to work with.

However, sometimes you do need to take it to the grinder again after heat treating it. Don't be afraid to give it another quick grind if the edge has been dulled in the process. If you do need to grind it again, simply give it another quick pass before taking it over to the sander. Otherwise, enjoy your new blade. You will still need to attach a handle, but that's it. You have a finished blade.

You may also want to make a sheath for it, but we'll get to that in chapter six.

Speaking of looking ahead, we're almost ready to turn our attention over to the heating and hardening process; but first, let's explore the more common knife grinds before concluding our discussion.

Types of Knife Grinds

Below, we are looking at the most common knife grinds. There exist more grinds than are listed here, but you will find these eight grinds on 99% of the knives that you use in your lifetime. Not only is considering these a good place for beginners to start, but these knife grinds are also a good place for beginners to end. You can be an expert and never branch out beyond these grinds; they're that wide-spread and common.

Asymmetrical: An asymmetrical grind is one that is, well, asymmetrical. But what exactly does this mean for the edge of your knife? To be asymmetrical, you must have one side that looks different from the other. We'll see that with the chisel grind, so that can't be the only feature. As it turns out, it isn't. The reason a chisel grind isn't an asymmetrical grind (despite being asymmetrical) is because one of the sides is left completely alone. To have an asymmetrical grind, you would need both sides of the blade to have a different grind. The absence of a grind is not itself, a grind. For this, you would need to have a convex grind and a compound grind, or any other combination of two grinds.

Chisel: A chisel grind has one side that is kept entirely flat. The spine of the blade stays straight and flat all the way to the point. The edge of the blade has a decently steep angle from the middle of the blade through to the point. This type of blade is not commonly used for knives themselves, but it is commonly found on chisels. Chisels might not be knives, but they are still blades. If you wanted a chisel point on a knife, you could absolutely do so, but this would greatly limit the useful applications you would have for that knife.

Compound Bevel: A compound bevel grind is a really cool looking grind that is more reminiscent of a bullet than a blade. This is achieved through the addition of a second bevel around the middle of the blade. The grind pushes back from the center of the edge for a bit, and then it hits the second bevel, which has a much less steeper angle. As mentioned, it looks like the front end of a bullet. This is an extremely common grind because it isn't necessarily a grind in and of itself. Almost any of the grind styles on this list could have a compound grind by simply adding a second bevel to the blade. The second bevel works to make cutting easier, so this grind should be among your earliest learned.

Convex: A convex grind uses a rounded curve to bring the two sides of the bevel together into an edge. This grind looks exactly like a V grind, only it isn't straight but curved. This curve makes this grind fantastic for cutting for pretty much the same reason the compound bevel grind does. The rounded nature of the blade also helps it stay sharp for longer when compared to blades that use straight lines coming toward the edge. You might think that these features would make it a widely used grind, but that actually isn't the case. It does stay sharper for a longer period than other blades, which is good because this shape has the adverse effect of making it harder to sharpen.

Flat (Full): The full flat grind should be the grind you master first. This is due to its simplicity. This grind doesn't use any curve or pull any tricks; it simply brings the two sides of the blade together in a straight line. We made a full flat grind earlier in the chapter. Because the two sides come together at such a sharp angle, it makes for an incredibly sharp edge, but with the trade-off that it is among the least durable grinds. The full flat only has a single bevel that reaches pretty much all the way back to the spine itself. It isn't used commonly outside of kitchen knives.

Flat (High): The high flat grind is extremely similar to the full flat, but it has a smaller edge. The edge of the full flat goes back all the way to the spine, but the high flat keeps a decent portion of the blade's cheek as thick as the spine before it then angles into the bevel. This is more commonly used than the full flat, though it has much the same effect when it comes to chopping and cutting.

Flat (Scandinavian or V Grind): Of the flat grinds, the Scandinavian, or V grind, is the most common grind. This grind follows the trend of the previous two, in that the edge starts much further down the cheek. The full flat is all edge, the high flat is 75% edge, and the Scandinavian grind is about 25% edge. This means that three-quarters of the blade is the same thickness as the spine. This makes for a heavier blade than the other two, but it is also more durable. One of the best aspects of this grind is that it is extremely easy to sharpen up. Of course it dulls easier too, so you'll find yourself sharpening it often.

Hollow: A hollow grind is often favored by hunters because it has an incredibly sharp edge that can be used for skinning animals. 75% of the blade is kept the same size as the spine before taking a convex curve to make the edge. The curve makes the blade look like the tip of an old-fashioned inking pen. Because the edge is so sharp and narrow, it is also incredibly easy to dull. This makes the hollow grind a bad choice for knives that will be used primarily for cutting or chopping. It isn't even all that good for stabbing, honestly, though it is the blade of choice for skinning animals or whittling wood.

BLADESMITHING

Chapter Summary

- Grinding your blade isn't very hard, but it is extremely important. Without grinding our blades, we would never have an edge to cut with.

- The first step to making a knife is to design it. This can be done by hand or online using a tool like Knifeprint.

- Once you have the design for your blade in place, you can roughly cut out the metal and then take it over to the grinder to grind out the shape. At this point, you will want to be extremely careful to hold the blade perfectly horizontal, so you get a smooth grind.

- Once you have a shaped blade, you need to find the center of the edge. This is done by applying some machining ink to the edge, then going over it with a drill bit. Flip the blade and go over it with a drill bit again to mark out your center line.

- With your center line marked, take the blade over to the grinder and make a very steep grind. We start steep like this, so we don't risk going over the center line. Once the first steep grind has been made, you can take a much less severe angle and start to grind along more of the cheek to create a nice, large edge. Keep grinding this way and changing the angle slightly to turn the edge from several bevels into one.

- Flip the knife over and grind out the other side.

- At this stage, the blade is sent off to be heat treated before being finished.

- You might need to grind the blade once again after it has been heat treated, but this isn't always the case. You will always have to sand it down using increasingly higher grit counts in order to create a beautiful and smooth finished product.

- There are eight grind styles that are found on 99% of the knives out there.

- An asymmetrical grind is one that uses two different styles of grinds, one for each side of the blade.

- A chisel grind has one side that is completely flat and another with a steep angle to create the look of a chisel. This grind isn't common on blades, but it is commonly found on, well, chisels.

- A compound bevel is a grind style that has more than one bevel.

- A convex grind has a rounded curve that brings both sides of the bevel into the edge.

- A full flat grind is the best grind for beginners to learn, as it is one of the easiest. This grind brings both sides of the blade together in a straight line.

- A high flat grind has a smaller edge than a full fall. The full flat edge takes up almost the whole of the blade, whereas a high flat takes up about 50% of the blade.

- A Scandinavian flat or V grind looks like the other flat grinds but only takes up 25% of the blade.

- A hollow grind keeps 75% of the blade the same size as the spine before bending into a convex curve.

In the next chapter, you will learn how to harden your knives. This is done by heating the blade to the right temperature in a forge first. Next, we would need to remove the blade and quench it in oil. Following that, we would have to heat it up again to temper it and make it strong. Finally, since we want to be able to hold our knives, we'll need to make an attractive handle for our freshly hardened blade.

CHAPTER FIVE

HEATING AND HARDENING

What do you think the most important part of knifemaking is? Did you say shaping the knife? Well, you'd be wrong. Grinding? Still nope. Although these are all absolutely important during the process, they pale in comparison to the heat treatment process. After all, we want to be able to cut and stab with our knives. If we try to cut with our knife before heat treating it, then it is likely to chip, warp, and distort. This is because the metal is not yet hard.

We often think that metal is just naturally hard—and to a degree this is true. However, most of the metal that we encounter in our daily lives has been heat treated and hardened. Metal is actually much softer before it is heat treated. If you want your knife to cut, then it must be hard enough to cut.

In this chapter, we will look at how we would heat treat our blades. We will be using a forge in our example, but you can use a torch or other heat source with a little bit of ingenuity. The key here is to get the blade to the right temperature to harden. To do this, we will prepare our forge, treat the blade, cool it off, temper it, then finish it. We technically looked at finishing in the last chapter, so we won't go over this step of the process again in depth; just enough to make our way back over to our sander for the finishing touches.

What Equipment You Need to Heat Treat Your Blades

If you have the ability to do your own heat treatments, then you'll need to have the right gear. A forge is obviously a great start, but it is far from the only gear you'll have to use. You may have purchased some of these tools after our discussion in chapter three, but it is worth going over them again briefly here again.

For safety reasons, you should have a first aid kit equipped with supplies for dealing with burns. Hopefully you won't burn yourself, but you want to be able to treat burns immediately if that does happen. To avoid getting burned, you should make sure to wear heat resistant gloves when working with your forge and heated blades. Tongs are also an important tool, which will allow you to place your blades into the forge and remove them without burning yourself. Remember that tongs get incredibly hot, so even while your hands are far away from the forge, it is still important to wear your gloves.

Your forge will need to be heated up some way. If you are using a gas- or fuel-powered forge, then you will need to make sure that you have gas and fuel on hand. We'll be looking at using a charcoal heated forge, which is the most traditional style. You'll want to get some charcoal, since it can get incredibly hot. You could also burn wood like a traditional fireplace if you wanted, as it will get the temperature to the right level, but you'll find that it could prove to be much more difficult because you'll need to keep adding more and more wood to the forge as it burns off.

One of the ways we increase the temperature of the forge quickly is to direct air into it. If you've ever seen a classical blacksmith's shop, then you've seen a bellows before. A bellows is a tool that is manually pumped to blow air into the forge to increase the temperature. After all, fire eats away the oxygen in the air, so you would need to supply it with fresh air to keep it burning as hot as possible. Anything that blows air—from a hair dryer to a leaf blower—will do the trick. In order to test the temperature of the steel, you'll also want a magnet, though you need to press it against the blade, and this can be difficult. If you can attach a magnet to the end of a metal rod, then you can use this to protect your hands.

The final thing we need is a container that won't break under high heat and oil for quenching our blades. Different bladesmiths have strong opinions on what the best type of oil to use is. Vegetable oil, olive oil, motor oil—all of these will work to cool off your blade. Just don't use water; water never works, and it will most likely damage your blade beyond repair. Stick with oil and get enough of it to fill up your container because you'll want to submerge the blade fully in it.

Finally, please keep in mind that this is a simple style of heat treating that is perfect for beginners still working with 1075 steel. It will work with pretty much any of the simple high carbon steels (those that begin with 10xx), but other techniques will be required for more complex steels. But I've recommended that beginners start with 1075 anyway, so this guide will focus more on 1075 than the others.

Step One: Preparing the Forge

If you don't have a forge yet, you can actually make one pretty easily. A makeshift forge only needs five bricks. Place them in a circle with an opening big enough for your air-source. This style of "forge" will allow you to heat treat your simple carbon steels without costing you more than $20-$30 for the entire process. This makeshift forge will be incredibly limited, but for those still at the beginner's stage of purchasing gear, this will do the trick; assuming that you have your forge purchased or built, the first step in heat treating your blades will be to heat up your forge.

This is done by filling the forge up with charcoal. Once the charcoal is in place, turn on your air source. This will ensure that it continues to get hotter and hotter rather than level out. Once the forge is packed and the air source is on, use a lighter, blowtorch, or whatever other fire source you have to heat the coals and get it going. Although many people think that fire is simply "hot," there are different levels of heat to a fire. We want to get the forge up to 2500°F, which is incredibly hot. This temperature is so high that to look at it directly would actually hurt your eyes. Trouble looking at the forge is one sign that you are getting close to this temperature, and another is when the coals turn from black to a white or light-grey color as they increase in heat.

It will take a little bit for the coals to heat up, though you don't need to apply the flame to them the entire time. Once they turn red, you can leave them be, as the air source will cause them to continue heating up. At this point, you can use your time best by pouring your oil into your heat-resistant container. You actually want to heat this up too because a room temperature quench could cause problems. This won't always be the case, but it is much more likely than with a heated quench. You can heat a piece of metal and put it into the oil to heat it, or you can put it on the stove for a while. The goal is to reach 130°F.

The forge should be hot enough by the time you finish preparing your quench.

Step Two: Heating the Blade

Now is the time to heat the blade, which will be the hardest part of using a makeshift forge. One of the reasons why you should invest in a real forge when you are ready is how it will distribute heat much more evenly. A makeshift forge is more haphazard with how the heat moves through it. The goal in heating a blade is to heat it up evenly, so not one section of the blade is hotter or cooler than the rest. The larger the blade, the harder it will be to distribute the heat evenly.

To begin heating the blade, all you need to do is use your tongs to place it in the forge. This is the easy part. It becomes a little more difficult again when you realize that you can actually overheat the blade. We're using a simple carbon steel, and you can burn the carbon right out of it if it is kept too hot for too long. Plus, if you let it heat up too much, it will straight up melt. Sometimes we want this, but not right now. To figure out if you are heating the blade up too much, we can use a rather simple technique: watch the color of the metal.

BLADESMITHING

For the purposes of harding a 10xx steel, we will want a temperature of just under 1500°F. You can tell your blade is getting there because it will turn red like a tomato. When you first put the blade into the forge, it will be the color of that specific metal. It will then start to glow as it heats up, and this red color will be the first real color it transitions to. We hit an interesting window here, which is when we will need to be extremely careful with our temperatures, but more on that in just a moment.

If you are interested in the color changes of carbon steel, here is a quick rundown. Around 1100°F, you can expect a burgundy shade of red. It is lighter but still distinctly burgundy around 1200°F. At 1300°F, the steel will start to turn more tomato color, with 1400°F and

1500°F seeing this color lighten up. At 1600°F, it goes from red to orange, then a lighter orange at 1700°F, and the lightest orange yet at 1800°F. 1900°F is so hot that the metal will begin to glow yellow, with the shade of yellow getting brighter at 2000°F and 2200°F. For our purposes, however, we will be staying well below the yellow range. Indeed, we won't even be messing with orange stages.

When our steel begins to turn tomato color, that's when we know it is reaching 1300°F. At this point, we will want to use our magnet. Steel actually loses its magnetism around 1350°F. So, once we see it turning red, we would then touch the magnet to it. If it still attracts, then it isn't hot enough. We want to make sure that we're testing the steel with the magnet at the moment it fails to stick. If our first attempt fails to leave the magnet sticking, then remove the blade from the flame for a second, so it can cool enough to attract the magnet again. Put the blade back into the forge and watch with the magnet. The moment the magnet takes, you know that you only need a little while longer. We want to heat it up to the critical temperature, which for carbon steel would be about 1475°F.

Another way to tell if the blade is hot enough is to put a little salt on it. Did you know that salt melts at 1474°F? If you sprinkle salt onto the blade and it melts, then you know it is hot enough. But there's a catch here. If you are using a makeshift forge, it will be difficult to ensure that the blade is heated evenly. A blade that is heated in some places but not in others isn't very useful. You don't want a blade that is hard in one spot and soft in another—we want the whole blade to be hard. The salt test can be a good way to see if the blade is heated evenly. If some of the salt doesn't melt, then you know that section of the blade needs to be heated further. Just be aware that carbon steel will melt around 2600°F to 2800°F.

Once the blade is heated evenly, it is then time to take it out of the fire and quench it.

Step Three: Quenching the Blade

Quenching the blade happens as quickly as possible after it reaches the critical temperature. You want the blade to be at critical temperature—or as close as possible—by the time it goes into the quench. Once the blade is at the critical temperature, grab it with the tongs and dunk it into the container of oil.

The oil will start to sizzle as it reacts to the heat. Keep a tight grip on the blade with the tongs and move it around; this will get rid of any air bubbles that formed from the dunking and the temperature. It will also have the added benefit of ensuring that the blade cools off evenly, since pretty much everything to do with heating and cooling the blade will be most effective when applied evenly.

When quenching your blade, you only want to dunk it into the oil once and keep it there. Moving it around is a positive, but pulling it out and dunking it in multiple times won't be. This is called an interrupted quench. This technique can be useful for some metals, but not for the 10xx simple carbon steel that we are using in this example. For our purposes, you will simply want to dunk the blade, give it a little bit of movement, then let it cool down.

You will know that the knife has hardened properly after a quick test; take out a file and try to scrape it against the knife. A hardened knife will be much tougher than a file, so you won't be able to scratch into it with the file. If you find that the file *does* scratch into the blade, then it has not been hardened properly. You can harden a blade a second time, however, so the good news is that your knife won't be ruined—it may just take you longer to finish it than you had previously thought.

Step Four: Tempering the Blade

So, we've just successfully hardened our blade. Congratulations. You probably think that's the end of the matter. After all, the blade is hard… isn't it?

Yup, it certainly is. Only… it's too hard. That might seem a weird thing to hear. How can a blade be too hard? But it is true—there is too much tension within the blade, so it is extremely easy to break it. If you happened to drop it, or if you tossed it at a wall or something, then it would probably shatter; just tiny, jagged pieces of knife going everywhere. This would be a nightmare. Plus, it's a horrible waste of all our time and effort.

Here's a way of thinking of it—have you ever made noodles? If not, then perhaps you've at least cooked them, so it should be easy enough to understand. You make noodles out of a dough. This dough in this step is soft, usually made of ingredients like egg, flour, and milk. These ingredients are all soft, so we would have to harden them into proper noodles. For noodles, this is done by drying them. Likewise, we would harden our blades by heating them. When the noodle is hard, you can grab it in your hands and break it in half with only a little bit of exerted pressure. Think of how stiff and brittle a spaghetti noodle is before you cook it—you can't bend it or do anything with it. This is what our knife is like currently. However, once you apply some heat to that spaghetti noodle, it will get all floppy, and it will bend, twirl, twist, and move in all sorts of ways that it previously couldn't before being heated. Although we don't want our blades to twist and turn, we do want them to be more flexible. The heating process hardened the blade to the point that it is now like a dry spaghetti noodle. Now, we need to heat it up again, so it doesn't break when cutting.

To do this, we will want to reach a much lower temperature than when we first hardened it. This process of re-heating a freshly hardened blade is called tempering. Our goal is to heat the blade to a temperature of 400°F. At this temperature, the steel will "relax" a bit, and the tension making the blade brittle will calm down.

Since the blade has been hardened already, the steel will change colors at different temperatures. At 420°F it will be yellow. This will slowly turn to orange, so by around the 450°F to 480°F range, it will be the color of a pumpkin. At 500°F, it will start to become a soft red, with a much darker red at 520°F. That red will quickly turn to burgundy at 540°F, then dark blue at 560°F and light blue at 600°F. However, we don't want to get there—we're looking to heat it to 400°F, which will be a golden brown color.

Achieving this color and temperature is much easier than hardening the blade in the first place. We don't need to get to such a crazy high temperature. In fact, you may have probably already noticed that 400°F is well within the temperature range of the oven in your kitchen. That's right—that's why we'll be baking your knife in the kitchen.

Set your oven to 400°F and put the knife inside. You can place it directly onto the middle rack. Set the timer for an hour. When it is finished, take it out and let it sit somewhere to cool off. You don't need to quench it again, since that is only for hardening. Once the knife has cooled, put it back into the oven again. When you take it out this time, pay close attention to the color. If it has turned that golden brown color, then it is good to go. If it hasn't yet, then you may want to put it in for another hour; just keep an eye on it this time and take it out if it changes colors before the hour is finished.

This is the easiest way to temper a blade, but it isn't the only way. Anything that will heat up the blade to a temperature of 400°F can be used. However, a kitchen oven will be extremely effective here because it applies heat naturally and evenly across the blade.

Regardless of how you do it, once you have that golden brown blade, you will then be ready for the next step.

Step Five: The Finishing Touches

Have noticed that your blade looks… well… kind of bad? The quenching process doesn't result in a nice looking blade; what you get instead is a blade that is covered in a weird substance called scale. If you've ever burnt a pizza before, then you may notice that it looks almost exactly like charred dough.

This is a perfectly natural part of the knifemaking process, even if it looks pretty disgusting. Many first-time bladesmiths pull their knives out of the quench, see them covered in scale, and assume that they did something wrong. Nope—this disgusting looking stuff is part of the process. Well, it is quite often part of the process. You can actually get lucky and quench a blade without the resulting scale. This is a more rare outcome than not, but it doesn't mean that anything went wrong; it just meant that you got lucky.

Don't worry about the scale on your blade—just take it over to the sander to clean it off. This will help you get a smooth blade, and it'll also remove that scale. We've already covered how to finish your blade off in the last chapter, so we won't go over it again here. Instead, let's take a quick look at adding a handle.

Step Six: Adding a Handle

A knife isn't very useful without a handle. You'll have a reduced grip, and this will result in more accidents, plus you won't be able to get the kind of power behind it that you would want to. A handle is what really makes a knife versatile. There are all sorts of ways to add a handle to your knife; one common way to do so is to drill holes into the blade, so you can stick the handle on either side and use rivets to hold it together.

But—oh no—we've already hardened our blade, and we didn't stop to think about the handle at all. There aren't any rivet holes to attach a handle!

Don't worry—this happens all the time, and it is far from the only way to attach a handle. Handles are a part of knifemaking that can get extremely complex. There are dozens of materials, styles, and looks that you can go for with your handle. Since we're just starting out, we won't want to get overly complicated, so we'll stick with a wood handle. Plus, we don't have any rivet holes, so we'll have to make due with what we've got.

You'll want to sand your blade prior to working on the handle, so don't forget that step. If you've polished your knife or anything like that since sanding it, then you'll want to sand it again because polishing can actually make it difficult to glue the handle into place when we do so in a few minutes. I recommend using a thin application of acetone over the handle because it will clean away any grit or grime that may be present. The cleaner the blade, the easier it will be to get the handle on properly.

Take your piece of wood—whatever type you feel like using—and lay your knife down on it. Take a pen or pencil and trace around it to get the shape for your handle. This will give you one side of your handle; don't forget that you need to do this step twice to get both sides. You can make your handle whatever size you want, but it is a good rule of thumb to trace it slightly larger than the knife. There are knives with handles that are smaller than the end of the blade, and you can see the metal poking out from the sides, but this type of handle could get uncomfortable to hold for long periods. If you err on the side of larger, then you can always sand it down to be smaller. Remember—it is always easier to take away materials than it is to add them.

Also, make sure that your knife is going along the wood grain. To go across the wood grain will result in a handle that is less structurally secure. With your handle traced out, it is now time to cut the wood. Use whatever tools you have available for cutting wood. You could even sand it down to size if that is what you prefer.

After you have cut your handle out, use a strong glue to stick it onto the knife. You want to glue both sides of the handle together at once. This step can be quite frustrating because if you don't manage to get the glue to stick properly then the handle will simply fall off down the road a little bit. It is because of how important this step is that we used the acetone to clean our blade. Once you have applied the glue and the two pieces of the handle you will want to use clamps or something similar to tightly squeeze both sides together. With your clamps in place, take a break. Go relax, get some sleep and come back twenty-four hours later.

You should have a handle that is held securely to your knife. Congratulations, you now have a handle on your newly crafted knife. But it probably isn't very good looking just yet. We only just cut out the rough shape needed for a handle after all. At this point you will want to head into the workshop and use the vice on your workbench to hold the blade securely. You want the vice to be holding the blade, not the handle, as you'll be shaping the handle. A wood rasp is a great tool for shaping the handle but you can use a file or even another knife, anything that will let you really work with and perfect the shape of the handle.

Once the shape is just right, you'll want to sand the handle down. We do this so the wood can be nice and smooth with no rises or bumps. Plus, we certainly don't want to get a splinter from this work. I recommend sanding the handle by hand rather than using an electric sander because wood doesn't need nearly as much sanding as metal does. Start with a 60 grit paper, then move up to a 120 grit paper, then a 200 grit paper. At this point, you may decide to stop, but I would continue next with a 300 grit paper and then a 400 grit paper. Doing so will ensure that the wood is extremely smooth, and it will help make sure that you can hold onto it without it slipping around in your hand.

The final step is to finish the handle. You may notice that the wood is rather boring-looking at this point. Most knives have a gorgeous wooden handle, but ours just kind of looks… plain. This is perfectly normal, and it isn't a sign that anything is wrong with the handle; it simply means that we haven't applied a wood finish. There are many, *many* kinds of wood finishes that we can use to finish our handles. Oil finishes are incredibly popular, but other options include varnish, staining, lacquer, or water-based finishes. Each of these finishes will give the wood a different look. You could even just paint the handle if you wanted, though any color besides white or blade has a tendency to look pretty silly.

Regardless of which finish you choose, this step has now brought you from designing your blade all the way through to heat treating it, quenching it, tempering it, and making a beautiful handle. That's what it takes to make a knife. Congratulations, and I hope many more are to come!

Remember that there are plenty of different styles you can make. There are different types of knives, bevel grinds, and handle styles that you can choose from. For our exercises, we have gone with the easiest and simplest of knives to use because it will make for a better introduction for beginners, and one that is less stressful. The best part about bladesmithing is trying to and succeeding in making a blade using a new style, technique or tool. So, get out there and get experimental with it!

BLADESMITHING

Chapter Summary

- Heat treating your blades doesn't have to be overly complicated. You can make your own forge with a few bricks, some coal, and a blowdryer.

- Start the forge by filling it with charcoal and lighting it. Use the blowdryer to blow air onto the coals, so the temperature rises up nice and high.

- Use your tongs to add the blade to the fire. We want the blade to reach a temperature of 1475°F, which is what we call the critical temperature. You can tell it is getting close to the critical temperature if a magnet will no longer stick to it. Another trick is to sprinkle some salt on the blade. If the salt melts, then you know the blade is at least 1474°F.

- Take the blade out of the forge and dunk it into a container of oil. This is called quenching the blade. Move the blade around inside the container, but don't take it out and dunk it again. This is a more complicated process and can cause some serious damage to the blade if you don't know what you're doing.

- Take a file and try to scrap the blade. If the file can't dig in, then you know the blade has hardened properly.

- However, the blade at this point is too hard, so it is incredibly brittle. Thus, we must temper the blade.

- Tempering the blade is basically cooking it again. This relieves the blade of a lot of the tension that made it too stiff.

- To temper the blade, simply set your kitchen oven to 400°F, then bake the blade for an hour. Let the blade cool off, and then bake it for another hour, or until it turns a golden brown color.

- Your blade should look kind of gross, as the quenching process often results in a layer of scale. This can be sanded off as part of your finishing step.

- The next step is creating a handle for your blade. You can do this in many ways, but one of the easiest ways to do so is to cut a piece of wood and use glue to attach it to the blade.

In the next chapter, you will learn how to create a beautiful sheath for your knife out of leather. This guide will walk you through cutting the leather, dying it, stitching it together, shaping it, and even making a belt loop, so you can easily take your knife out with you into the wild.

CHAPTER SIX

CREATING A SHEATH

We now have a brand new knife with a beautiful handle, and you could stop here if you wanted. You have everything you need to continue making all sorts of knives. However, after a while, you'll probably find that your knife is still missing something. It's sharp, beautifully shaped, and hard enough to cut or stab. But what about storage?

If you've added a lanyard hole to your blade's hilt, then you can attach it to a belt. But how do you stop it from stabbing your leg? If it is freshly grinded into a razor-sharp edge, then you're definitely going to want some way of keeping it safe. You can always stick it in a drawer or on a shelf, but then what use is that? You can't really use that knife for much unless you're at home. If you just stuffed it in a backpack or something, then you'd have to wrap it up with a towel or similar instrument to ensure it doesn't damage anything else you were carrying.

Simply put—if we want our knives to be practical, then we need to also know how to make a sheath so we can carry them easily.

In this chapter, we will be doing just that. For this project, we're going to make a leather sheath, so you will need to purchase some leather. Other important tools include a ruler, a hammer, some thread and needles, a hole punch, something to cut the leather with, glue, sandpaper, and a pair of pliers. As with any project, you shouldn't feel compelled to follow this guide to the T; for example, we'll discuss dying our sheath, but there is no reason you should feel compelled to do so. We'll also be looking at a pretty simple sheath—one that looks beautiful but with a very simple design, yet you could still spend hours adding designs and details, such as etchings or raised thread designs to the sheath if you wanted to. This guide will merely show you how this process is done; it isn't meant to argue that one style is any better or worse than another.

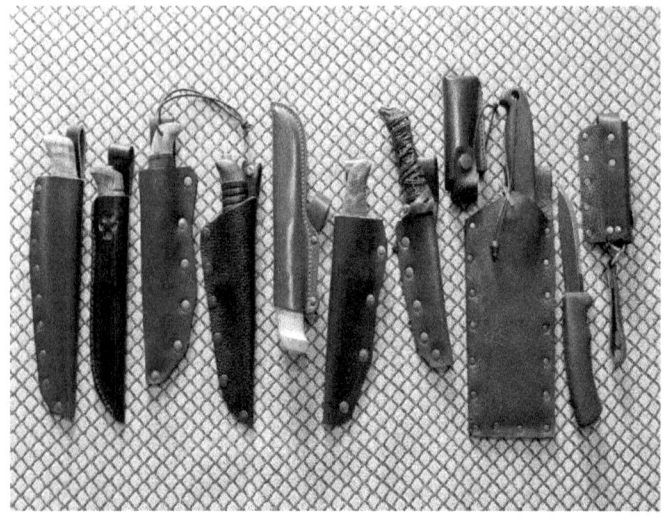

Step 1: Preparing the Shape

Once you've purchased a piece of leather, you've probably noticed that it isn't particularly shaped up to look like anything. We want this to be a sheath for a knife, so we'll have to shape it ourselves. Rather than cut into the leather to begin with, pull out your pen and paper once again.

Set your knife down on the paper and trace around it. This will result in a fairly detailed representation of your knife. However, if we were to use only this, it wouldn't give us enough space. We can know by tracing a knife that we can pretend is inside the sheath. Sticking with your pen and paper, add lines to the drawing to create the sheath itself. This will result in a strong line that represents your knife and another larger line that represents the sheath. When drawing your knife, make sure to draw the handle, too. We'll use this guide to create a belt loop later.

Use a pair of scissors to cut out your drawing. Take your piece of leather and fold it in half. Place the drawing onto the leather as close to the edge as possible, and then use your leather cutting tool to cut it out. This should result in a piece of leather with two sides in the shape of your design.

Step 2: Dying the Leather

This is a quick step. Simply dye the leather with whatever dye you are planning to use for the sheath. At this point, the color will be quite mild, but we'll be adding to it later on to make it stand out.

The reason that we dye the leather this early is simply because it is the easiest time to do so. We haven't folded in, glued it, drilled into it, or done anything to alter it. This means that we can reach every piece of the leather without any issues.

Step 3: Etch Out a Groove

Pick one side of the leather to be outside and one to be the inside. Using your leather cutting tool, etch out a grove carefully all along the edge of the knife, a few millimeters from the edge. At this point, we won't be doing much of anything with the groove. Later, however, this will be used to keep our stitching sunken and out of the way.

If we skip this step, then, when we later sew together the two sides of the sheath, we'll find that the stitching is raised, which will have the adverse effect of making it more likely to catch on something or come undone.

Step 4: Shape, Cut, and Glue on the Edge's Bottom

Set your leather down on a piece of paper and trace along one side. This will give you a line that follows the flowing shape of the sheath. Remove the leather and then add a straight line about a millimeter tall to both ends of this line. Following the same flow again, draw a second line that parallels the first and connects the two straight lines together. This will make the middle section of the sheath. When the knife goes into the sheath, this piece will be against the blade to help protect the sheather. Use your scissors to cut it out.

Take your sketch of the middle section and place it onto the leather you have cut for the sheath. Lightly trace along the edge of the piece of paper, so you mark how much space it takes up from the edge of the sheath. You want this to be on the inside of the sheath because it shows us where we will be gluing in a moment. However, before you worry about that, take your sketch and use it to cut out another piece of leather. Don't double up the leather for this cut like you did originally.

You now have a piece of leather that will serve as the middle part of your sheath. Take your sandpaper and lightly sand down the section you've marked for gluing. Apply your glue and give it a few minutes to dry, after which you will press the middle piece firmly into place.

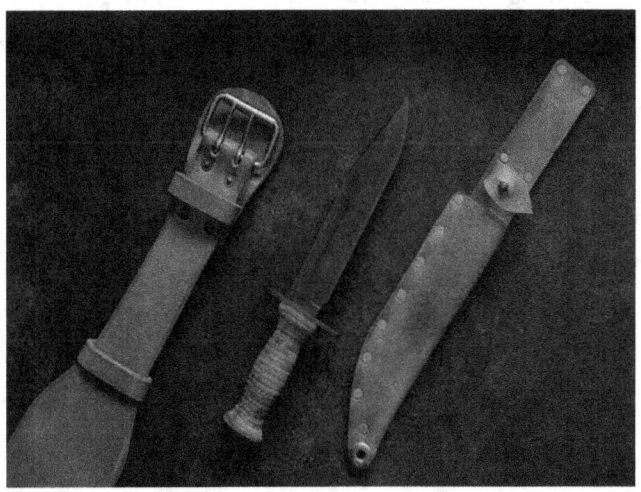

Step 5: Make the Belt Loop

The piece of leather you've cut for the sheath should have a bar jutting from the top that looks like the handle of your knife. This will be for the belt loop. Dip this piece of the leather into some water for ten to twenty seconds, so it will bend properly.

Grab a belt if you have one and place it against the leather. Then, bend the belt loop over the belt and press it against the sheath; doing this will show you how long the belt loop is. You don't want it to be too long—only two or three millimeters past the end of the belt. Mark this section, then cut the loop to size. You can keep the loop rectangular or you can shape it to be round, triangular, or any other style as taste dictates.

Having marked where the loop connects to the sheath, sand this section and apply glue. Once the glue has dried, press down on the belt handle firmly. Dying the edges of the loop can be difficult to do after it has been glued into place, so while the glue is drying is a great time to do so if you are planning to dye the sheath.

Once the belt loop is glued into place, use your cutting tool to etch out grooves for stitching. Next, take a hole punch, sharp fork, or tiny screwdriver and make evenly spaced holes along the etching. Take your thread and poke the needle through these holes, then stitch the loop into place. Tie the thread off afterwards. If done properly, the belt loop should now be in place and the stitching sunken and out of the way.

Step 6: Fold, Mark, Drill, Groove, and Stitch the Sheath Together

Apply glue to the middle piece that you recently added. Also, glue entirely along the opposite side. In addition, you should apply glue in a decently thick line straight down the middle of the sheath. Again—let the glue dry first before folding the sheath in the middle.

Once folded, you will notice that the piece of leather is starting to finally resemble the sheath it will be when it is finished. With the glue dried and the two pieces together, go along the line you etched in step three and drill or poke evenly spaced holes along it. It can be easier to use a very small drill bit to make the holes, but you should mark them prior to drilling.

Flip the sheath over and use the holes to place a game of connect the dots. Use your cutting tool to etch another line. The holes should be evenly spaced and made according to the etched groove from the other side, thus resulting in a nearly perfect recreation of the original etched line. This is, again, to sink down the stitching, which will be the final part of this step. Sew along this line using the holes to hold the bottom of the sheath firmly in place.

Step 7: Cut Off Excess

At this point, you can say that you *almost* have a sheath. There are still a few steps left, but most of them are for finalizing and decorating the sheath. However, chances are good that your original cut leather wasn't the perfect shape. Plus, you etched the original groove into the leather a few millimeters from the edge, so there should be some excess leather.

Take your cutting tool and carefully trim this away. As you cut away the leather you will also want to sand the edges to make them nice and smooth; otherwise, the sheath will have the appearance of a low quality creation. You may also want to chamfer the edges of the sheath to give it a more appealing look.

Step 8: Dye, Burn or Decorate

This is a good time to apply another coat of wax. The edges of the sheath have gone from being as thick as one piece of leather to being as thick as three pieces, so dying the edges is much easier to do now than before.

One technique that can really give your sheath an amazing appearance is to use a burn stick along the edges. This is a type of tool that is made out of a wooden stick. Rubbing a burn stick along the leather can really quickly cause a lot of friction that raises the temperature of the leather and burns it slightly, giving it an older appearance than it really is. It also tends to really bring the whole piece together, so I highly recommend this step.

If you want to add decorations to the sheath, now is the time. Lightly wet the leather, so that it is more pliable, and use your cutting tool to work in designs, shapes, patterns, images, or you could even simply etch another line or two along the edge to give it a more well-defined look.

Step 9: Form the Final Shape

The sheath is almost done, but now we want to make sure that it is shaped perfectly for our knife. First, take the knife and wrap it in some tin foil or a few layers of plastic wrap. This is done for three reasons—the first is that we don't want the knife damaging the leather during this step because it will be quite fragile still. The other effect it has is to ensure that the knife isn't held in place too tightly. We want to make sure that it stays, but a tiny bit of wiggle room will make it much easier to sheath and unsheath the blade. The final reason is that we'll be soaking the sheath, and we don't want to get the blade wet.

Dunk your sheath in water for ten to twenty seconds to wet it thoroughly. Once wet, insert your blade into the sheath, so that the blade is facing down toward the middle piece of leather we added in step four.

Use your hand or another object to push the ends of the sheath tightly together around the edge of the knife. This will help ensure that the knife is secure when it is in the sheath, and it will result in a much more uniquely shaped sheath too; one that is specifically designed for this particular blade. Doing this can take your creation from being a simple sheath to a part of the knife, which, in turn, raises the value of that particular blade as a whole.

Leave the sheath to dry with the knife in it. Once it is dry, you can remove the knife and take off the plastic wrap. You should notice something cool here. Even though you've removed the knife, the sheath should still be shaped perfectly, as if the knife was inside of it. This is because dried leather holds its shape extremely well, so, by soaking it to reshape it and letting it dry, we've created a perfectly sized sheath.

Step 10: Finish Decorating

At this point, the color of the dry should still be pretty lackluster. It is certainly a different color from the original leather, but it doesn't have very much pop. To give the sheath a shine, we can use an application of special wax.

Purchase both wax and linseed oil, then mix the two together in a pan at a 1-1 ratio. Heat up the pan, so that the two ingredients blend together, then take it off the heat and let it cool back down. Once cool, apply the wax to the sheath in heavy layers. This isn't like painting, so don't worry if it is globby and not spread evenly. Use a blowdryer or another heat source that is easy to direct to melt the wax over the sheath. This should result in a much nicer looking color, though you can always repeat this process multiple times, until it looks just the way you want it to.

Assuming that your first application is successful, all you need to do now is buff the sheath. This will even out the wax and remove any excess that hasn't stained the leather, and it will result in a beautiful and shiny sheath. If your first application of wax wasn't enough, then remember to buff the sheath before the next application.

Congratulations, you now have a beautiful sheath for your brand new knife. By taking this process and changing different elements you can create endless amounts of unique sheaths. Add decorations, change the color, the choice is entirely up to you.

Chapter Summary

- The first step in making a sheath is to prepare the shape. Place your knife on a sheet of paper and trace around it. Cut out the tracing and place it on a piece of leather that has been folded in half. Then, cut out the leather.

- After cutting out the letter, it is the easiest time to dye it because it is simply flat during this stage.

- Use a leather cutting tool to etch a groove into the leather along the edge. This will mark the location where the stitching goes, and the reason we etch a groove is so the stitching will be sunken into the leather and harder to untie accidentally.

- Next, you would cut out a piece of leather that follows the lines of the sheath. This small piece is used to create the center of the sheath: the part upon which the knife's blade will rest. This is then glued onto the bottom of the leather.

- The piece of leather that comes from tracing the handle of the knife is used to create a belt loop. The leather is briefly soaked in water to make it flexible. It is then folded down, and the location it connects to the sheath is marked, and it is glued on. After that, you etch a groove, poke holes along the groove, and then use those holes for sewing the belt loop into place.

- Apply glue to the ends of the leather and to the middle, then fold it in half. Use a drill and create evenly spaced holes along the groove. Flip the sheath over and use the holes you just drilled to guide your cutting tool as you etch out an identical groove on the back of the sheath. Again, these grooves hide the stitches you sew into place now to hold both sides together.

- At this point, the sheath is almost ready, but you need to cut away any excess leather.

- Dye the sheath along the edges. A burn stick can be used along the edges to create a rich appearance. This is a good time for any other decorating you want to do, such as adding more grooves to the sheath or creating a pattern.

- Take your knife and wrap it in plastic. Then, soak the sheath in water for ten to twenty seconds, and then stick the knife into it. Press the sheath down as tightly as you can against the blade. This is incredibly easy to do when wet, but as the sheath dries, it will harden up. After it is dry, you can then remove your knife without losing the shape of the sheath.

- All that remains is any additional decorating you want to do. To really bring out the color, you should use a mixture of one part linseed oil and one part wax that has been heated on a flame. Apply this mixture generously over the sheath and use a blowdryer to melt it, so it coats the sheath evenly. This will bring out the color of the dye and really make it shine.

In the next chapter, you will learn how to maintain your knives, so they can stay in excellent condition for as long as possible. These tricks and techniques are incredibly easy, but they are important for caring for your blade properly, like it deserves.

CHAPTER SEVEN

MAINTAINING YOUR KNIVES

Making a sheath is only one of the ways that we would keep our knives in top condition, but it is far from the only one. Since we want our blades to last us as long as possible, it falls on us to take steps to ensure that they aren't exposed to unnecessary damage. It is also our responsibility to care for our blades, especially those made of carbon steel, as they have their own special needs when compared to other metals.

In this chapter, we'll look at how to care for a blade generally, care for a carbon steel blade specifically, and the steps needed to be taken to maintain our blades. These steps aren't very difficult; in fact, they are exceedingly easy. The difficult part is remembering to keep up this care and maintain activities.

Keep It Dry; Only Hand Wash

Your knife is made of metal, and metal will rust when exposed to water if it hasn't been properly treated for it. We don't treat our blades for water protection because it simply isn't important when it comes to making knives.

However, our knives certainly can get dirty. Say you use it to cut some strawberries—the inside of the strawberry will stain the blade red, making it look almost like a murder weapon. Since we don't want people to get the wrong idea, it is best to wash the blade. This does mean getting it wet, of course.

Washing your blade is perfectly fine, though you should do it by hand. The important thing about washing a blade is drying it afterwards. Use a towel that will absorb the water and not just push it around on the blade. You want to ensure that the blade is dry as soon as possible after washing.

Never Put Your Knives in the Dishwasher

As mentioned above, it is better to handwash your blade. In fact, you should *only* hand wash your blade. One effect this has is that it limits the amount of time that the blade is wet. If you put the blade into a dishwasher, then it has to be in there while it runs through the entire cleaning cycle, and this can take anywhere from one to three hours.

There is also an even bigger problem with dishwashing a knife. A dishwasher tends to be much more aggressive than hand washing techniques. It uses jets and it tosses the water around all throughout the machine quite forcefully, which can cause damage to the blade, especially the handle.

Another issue is that a dishwasher is also far more likely to reach higher temperatures when compared to washing by hand, and this can cause issues with the blade if temperatures reach too high. Chances are that it won't be a problem, but there have certainly been blades ruined through this practice.

You may think that a dishwasher is the easiest way to deep clean your blade, but you'd be surprised at how effective some soapy water and a dish rag can be.

Keep It Protected in a Sheath or Wrap

Keeping your blade in a sheath or wrap is always the best idea for storage. The sheath will protect the blade from the elements, so you won't need to worry about it getting wet. Even more importantly, a sheath will keep the edge of the blade nice and sharp, as it is a way of gently holding the blade in place.

If you are taking the blade out with you anywhere, make sure you keep it sheathed properly in the leather sheath you made in the previous chapter.

Clean the Blade After Cutting Anything Acidic

This is actually more so a tip for carbon steel blades than blades themselves. Carbon steel has a strong reaction to acid, and if it is left untended to, then it will actually start to corrode around the metal of the blade. While water will rust a blade, such an incident pales in comparison to how much and how quickly acid will cause it to rust.

You may be thinking that you don't need to worry about this; after all, who interacts with acid regularly? Well, actually, most of us. If you've had a tomato, orange, lime, lemon, or onion today, then you've interacted with acid. These fruits and vegetables are highly acidic, so cutting through them with your knife thus exposes it to acid.

Don't worry though—it's not as bad as you might think. Although acid will make the blade corrode pretty quickly, it isn't the kind of thing that will happen in front of your eyes. You have no reason to be scared to cut through any of these acidic products. Simply make your cut, then take your blade over to the sink and wash it by hand with some soapy water. This will remove any of the acid, so that it doesn't stay on the blade long enough to damage it.

If you don't have the time to wash the blade by hand—such as when you're in the middle of cooking dinner, then rinsing it off will work. Just use some warm water to wash the acid from the blade, then wash it with soapy water as soon as you get a chance.

Oil the Blade After Cleaning

This is another tip for carbon steel blades. An oil, like camellia oil, should be applied to the blade when you finish washing. Applying oil works to create a protective layer over the blade. This won't completely nullify the effects of acid and water, but it will help protect it against both of them.

To oil your blade, apply a couple drops onto the cheek. Although a rag is the easiest way to spread the oil around, your fingers will be able to do the trick just as well. You will want a nice, even layer of oil, just like how we wanted to heat our blade evenly during the hardening phase. If you are using your fingers to oil the blade then you should avoid the edge. Get the oil as close to the edge as possible, but don't run your fingers across it. This is safety 101, and it shouldn't even need to be said, but it is always better to be safe than sorry (and lose a finger!).

Sharpen the Grind Yearly (at the least)

Your knife will dull over time. It happens. We'll sharpen it as we need, but this type of maintenance is really a matter of delaying the enviable. Your blade will eventually dull to the point where it just doesn't hold an edge the way it used to.

This is a normal process, though also a disappointing one. The best way to get around this issue is to take your blade to the grinder once a year. Doing so will hone up the edge nicely and ensure that it is always in the best shape.

Sharpen Your Blade as Needed

Between your yearly grinds, the blade will still need to be sharpened from time to time. This is less care and more maintenance. You are maintaining the sharpness of the blade; not the edge itself. You would grind out the edge to maintain it; this is simply about making sure that it cuts.

First, get yourself a honing rod. This isn't actually used for sharpening itself, but it will remove any small, impossible-to-see bits of steel that are jutting out and reducing the cutting edge of the blade.

Next, you will want to get something to sharpen it with, like a whetstone or a knife sharpener. A whetstone is the classic way of sharpening a blade, and it is simply a piece of stone that you would run against the edge to clean it up. A knife sharpener, on the other hand, is a more modern invention. These typically look like small pieces of plastic with a slit through which you would run the blade. Inside the slit is another piece—mostly out of sight—that is tough enough to sharpen the edge.

Remember that this approach won't make your knives last forever. They'll last longer, but they'll still wear and tear over time. We want our knives to cut, so they would not be good to us when they are dull.

Chapter Summary

- Regular care and maintenance is required if you want your knife to stay in top quality for as long as possible.

- It is important to keep your blade dry, as it is a metal that will rust.

- But we still have to clean our blades, so do it by hand and dry it off quickly afterwards.

- Don't wash your knives in a dishwasher; a dishwasher will keep the blade wet for longer than it should be. It is also rougher than washing by hand, and the higher temperature can cause issues in rare cases.

- Keeping your knife in the sheath will prevent it from getting wet; plus, it will prevent the knife from accidentally stabbing or cutting anything while being transported or stored.

- Carbon steel has a negative reaction when exposed to acid: it begins to corrode. Anytime you cut highly acidic foods—like citrus fruit or tomatoes and onions—you should then immediately wash the blade. If you can't wash it, then at least rinse it off to remove the acid.

- Oiling your blade will create a protective barrier that helps reduce the damaging effects of water or acid. Oil won't prevent these elements from damaging the blade; it will merely reduce how quickly the blade is affected negatively.

- You will want to take the knife to the grinder once a year to ensure that the edge stays nice and sharp over the long run.

- In the short term, you can use a whetstone or a knife sharpener to sharpen up the edge as needed.

- A honing rod won't actually sharpen a knife, but it will remove any tiny pieces of steel that are sticking out to reduce the overall sharpness of the blade.

FINAL WORDS

We've now reached the last part of our journey together. It has been exciting to take you from discovering bladesmithing, all the way through to creating your very first blade. Bladesmithing is an exhilarating skill to practice. You must be careful, however, because it deals with deadly tools, but that will keep your attention narrowed down on the project at hand, and there is so much variation between blades that you could work for years at this skill and never make the same blade twice.

On the other hand, that is very unlikely to happen. Most people begin trying a few different styles before finding the ones that they like the most and narrowing down their practice to master that particular design. You might find yourself doing just this. I do hope that you will continue to experiment and try new styles as you go forward with this skill because it is this aspect of bladesmithing that is the most enjoyable to me, and it would fill my heart with joy to be able to share it with you.

As you grow and continue, don't be afraid to come back and reread sections of this book as you need it. It should serve as a guide for helping you to grow from beginner to expert. To that end, let us quickly recap what we've covered in the pages above.

In chapter one, we looked at how to get started with this awesome skill. To achieve this, we studied how to pick the right space for our bladesmithing workshop, including the health concerns we have to take into consideration while scoping out a suitable location. From there, we looked at how to layout the space once we had it, and the safety tips we should implement in our workshops to minimize risk of harm.

Chapter two was all about designing your knife. We started by looking at the anatomy of the knife, so we could learn what each piece was called and what purpose it served. This then led us into a discussion about the most popular knife designs in use today.

Chapter three looked at the tools we use to make our knives. These can grow to be quite expensive; far more so than the average beginner is willing to pay to try out a new skill, so instead of just jumping into the deep end, we divided the gear into three categories (beginner, intermediate, and expert), so we can grow our workshop and tools naturally. We also looked at the extra gear you may want to invest in further down the road.

Chapter four was all about grinding. Grinding is an important part of making a blade because you can't have an edge without a grind. We looked at how to use a belt grinder to create our edge, but this technique can be applied to grinding knives in general.

As sharp as your knife may be after grinding, it isn't useful until it has been hardened. To learn how to do this, chapter five looked at the equipment you need to heat treat a knife. Then, this was followed by a step-by-step guide on how it is done, and, finally, a look at the most common knife grinds of today.

Once you have a hardened knife, you are pretty much good to go. However, most people would like to have a way to carry and store their knives safely. To achieve this, we looked at a step-by-step guide on creating our own leather sheaths in chapter six.

Finally, we closed out the book with chapter seven's discussion on how to maintain your knives. It is one thing to make a knife and another to maintain it. You don't want to have to make a new knife every other year simply because you weren't careful enough with the blade.

Going forward from here, there are many different areas in which you could choose to expand. You could try out a new knife design, make the same design but with a different grind, or use a different type of metal beyond 10xx steel, to name a few. Although you may have thought all blades were pretty much the same when you started the book, you know now that there is a whole world of possibility waiting to be explored.

So, get out there and try something new. If it fails, then try it again. If it succeeds, then celebrate and use your new knife to cut yourself a piece of cake.

www.ingramcontent.com/pod-product-compliance
Lightning Source LLC
Chambersburg PA
CBHW050321120526
44592CB00014B/2000